2-D Geometry

Picturing Polygons

S0-AGI-700

Grade 5

Also appropriate for Grade 6

Doug Clements
Cornelia Tierney
Megan Murray
Joan Akers
Julie Sarama

Developed at TERC, Cambridge, Massachusetts

Dale Seymour Publications®
White Plains, New York

The *Investigations* curriculum was developed at TERC (formerly Technical Education Research Centers) in collaboration with Kent State University and the State University of New York at Buffalo. The work was supported in part by National Science Foundation Grant No. ESI-9050210. TERC is a nonprofit company working to improve mathematics and science education. TERC is located at 2067 Massachusetts Avenue, Cambridge, MA 02140.

This project was supported, in part, by the
National Science Foundation
Opinions expressed are those of the authors and not necessarily those of the Foundation

Managing Editor: Catherine Anderson
Series Editor: Beverly Cory
Manuscript Editor: Nancy Tune
Revision Team: Laura Marshall Alavosus, Ellen Harding, Patty Green Holubar, Suzanne Knott, Beverly Hersh Lozoff
ESL Consultant: Nancy Sokol Green
Production/Manufacturing Director: Janet Yearian
Production/Manufacturing Supervisor: Karen Edmonds
Production/Manufacturing Coordinator: Barbara Atmore
Design Manager: Jeff Kelly
Design: Don Taka
Illustrations: DJ Simison, Carl Yoshihara
Cover: Bay Graphics
Composition: Archetype Book Composition

This book is published by Dale Seymour Publications®, an imprint of Addison Wesley Longman, Inc.

Dale Seymour Publications
10 Bank Street
White Plains, NY 10602
Customer Serivce: 1-800-872-1100

Order number DS47044
ISBN 1-57232-797-9
5 6 7 8 9 10-ML-02

Printed on Recycled Paper

TERC

Principal Investigator Susan Jo Russell

Co-Principal Investigator Cornelia Tierney

Director of Research and Evaluation Jan Mokros

Curriculum Development
Joan Akers
Michael T. Battista
Mary Berle-Carman
Douglas H. Clements
Karen Economopoulos
Claryce Evans
Marlene Kliman
Cliff Konold
Jan Mokros
Megan Murray
Ricardo Nemirovsky
Tracy Noble
Andee Rubin
Susan Jo Russell
Margie Singer
Cornelia Tierney

Evaluation and Assessment
Mary Berle-Carman
Jan Mokros
Andee Rubin
Tracey Wright

Teacher Support
Kabba Colley
Karen Economopoulos
Anne Goodrow
Nancy Ishihara
Liana Laughlin
Jerrie Moffett
Megan Murray
Margie Singer
Dewi Win
Virginia Woolley
Tracey Wright
Lisa Yaffee

Administration and Production
Irene Baker
Amy Catlin
Amy Taber

**Cooperating Classrooms
for This Unit**
Sarah Novogrodsky
Jill Berg
*Cambridge Public Schools
Cambridge, MA*

Philip Bronstein
Janis Held
*Transit Middle School
Williamsville, NY*

Technology Development
Douglas H. Clements
Julie Sarama

Video Production
David A. Smith
Judy Storeygard

Consultants and Advisors
Deborah Lowenberg Ball
Marilyn Burns
Mary Johnson
James J. Kaput
Mary M. Lindquist
Leslie P. Steffe
Grayson Wheatley

Graduate Assistants
Richard Aistrope
Kathryn Battista
Caroline Borrow
William Hunt
Kent State University

Jeffrey Barrett
Julie Sarama
Sudha Swaminathan
Elaine Vukelic
State University of New York at Buffalo

Dan Gillette
Irene Hall
Harvard Graduate School of Education

Revisions and Home Materials
Cathy Miles Grant
Marlene Kliman
Margaret McGaffigan
Megan Murray
Kim O'Neil
Andee Rubin
Susan Jo Russell
Lisa Seyferth
Myriam Steinback
Judy Storeygard
Anna Suarez
Cornelia Tierney
Carol Walker
Tracey Wright

CONTENTS

WHERE TO START

The first-time user of *Picturing Polygons* should read the following:

When you next teach this same unit, you can begin to read more of the background. Each time you present the unit, you will learn more about how your students understand the mathematical ideas.

Investigations in Number, Data, and Space® is a K–5 mathematics curriculum with four major goals:

- to offer students meaningful mathematical problems
- to emphasize depth in mathematical thinking rather than superficial exposure to a series of fragmented topics
- to communicate mathematics content and pedagogy to teachers
- to substantially expand the pool of mathematically literate students

The *Investigations* curriculum embodies a new approach based on years of research about how children learn mathematics. Each grade level consists of a set of separate units, each offering 2–8 weeks of work. These units of study are presented through investigations that involve students in the exploration of major mathematical ideas.

Approaching the mathematics content through investigations helps students develop flexibility and confidence in approaching problems, fluency in using mathematical skills and tools to solve problems, and proficiency in evaluating their solutions. Students also build a repertoire of ways to communicate about their mathematical thinking, while their enjoyment and appreciation of mathematics grows.

The investigations are carefully designed to invite all students into mathematics—girls and boys, members of diverse cultural, ethnic, and language groups, and students with different strengths and interests. Problem contexts often call on students to share experiences from their family, culture, or community. The curriculum eliminates barriers—such as work in isolation from peers, or emphasis on speed and memorization—that exclude some students from participating successfully in mathematics. The following aspects of the curriculum ensure that all students are included in significant mathematics learning:

- Students spend time exploring problems in depth.
- They find more than one solution to many of the problems they work on.

- They invent their own strategies and approaches, rather than rely on memorized procedures.
- They choose from a variety of concrete materials and appropriate technology, including calculators, as a natural part of their everyday mathematical work.
- They express their mathematical thinking through drawing, writing, and talking.
- They work in a variety of groupings—as a whole class, individually, in pairs, and in small groups.
- They move around the classroom as they explore the mathematics in their environment and talk with their peers.

While reading and other language activities are typically given a great deal of time and emphasis in elementary classrooms, mathematics often does not get the time it needs. If students are to experience mathematics in depth, they must have enough time to become engaged in real mathematical problems. We believe that a minimum of 5 hours of mathematics classroom time a week—about an hour a day—is critical at the elementary level. The scope and pacing of the *Investigations* curriculum are based on that belief.

We explain more about the pedagogy and principles that underlie these investigations in Teacher Notes throughout the units. For correlations of the curriculum to the NCTM Standards and further help in using this research-based program for teaching mathematics, see the following books, available from Dale Seymour Publications:

- *Implementing the* Investigations in Number, Data, and Space® *Curriculum*
- *Beyond Arithmetic: Changing Mathematics in the Elementary Classroom* by Jan Mokros, Susan Jo Russell, and Karen Economopoulos

This book is one of the curriculum units for *Investigations in Number, Data, and Space.* In addition to providing part of a complete mathematics curriculum for your students, this unit offers information to support your own professional development. You, the teacher, are the person who will make this curriculum come alive in the classroom; the book for each unit is your main support system.

Although the curriculum does not include student textbooks, reproducible sheets for student work are provided in the unit and are also available as Student Activity Booklets. Students work actively with objects and experiences in their own environment and with a variety of manipulative materials and technology, rather than with a book of instruction and problems. We strongly recommend use of the overhead projector as a way to present problems, to focus group discussion, and to help students share ideas and strategies.

Ultimately, every teacher will use these investigations in ways that make sense for his or her particular style, the particular group of students, and the constraints and supports of a particular school environment. Each unit offers information and guidance for a wide variety of situations, drawn from our collaborations with many teachers and students over many years. Our goal in this book is to help you, a professional educator, implement this curriculum in a way that will give all your students access to mathematical power.

Investigation Format

The opening two pages of each investigation help you get ready for the work that follows.

What Happens This gives a synopsis of each session or block of sessions.

Mathematical Emphasis This lists the most important ideas and processes students will encounter in this investigation.

What to Plan Ahead of Time These lists alert you to materials to gather, sheets to duplicate, transparencies to make, and anything else you need to do before starting.

INVESTIGATION 2

Triangles and Quadrilaterals

What Happens

Sessions 1, 2, and 3: Sorting Polygons After a brief homework review, students begin sorting polygons: the three-sided figures and four-sided figures from the Guess My Rule Cards. They think of different ways to categorize triangles and quadrilaterals and then complete the statements "All triangles...," "Some triangles...," "All quadrilaterals...," and "Some quadrilaterals...." Finally they play Guess My Rule with all the cards.

Sessions 4 and 5: Making Shapes That Follow Rules Using Power Polygons off the computer and *Geo-Logo*'s coordinate commands jumpto and setxy on the computer, students make triangles and quadrilaterals that fit given descriptions. They then discuss which of the shapes are impossible to make and which are difficult to make.

Sessions 6 and 7: Using Move and Turn Commands Students work with *Geo-Logo* move commands (fd and bk) and turn commands (rt and lt). They write procedures that draw a square and a rectangle, and see how they can use the repeat command to write the same procedures in a shorter form. On computer, students use *Geo-Logo*'s move and turn commands to draw an equilateral triangle (and, if time permits, other shapes). In an off-computer assessment, students answer questions about the hierarchical categories of polygons; for example, why a square is also a rhombus. As a follow-up to the computer work, students discuss the difference between turns and angles.

Session 8: Finding Angle Sizes Students use a number of different strategies to find the sizes of angles in the Power Polygons. They discuss their strategies, and they learn to use Turtle Turners as a way of checking their accuracy. For homework, students draw angles by estimating the size, then check each one with a Turtle Turner.

Session 9: Angles and Turns Together Students play a game on the computer that relates the size of an angle to the size of the turn made to form that angle (the supplement of the angle). Off the computer, they estimate and draw some of the angles in the Power Polygons sets, and they use Turtle Turners to measure turns and angles in triangles.

Mathematical Emphasis

- Reasoning and communicating about properties of geometric shapes
- Sorting and classifying triangles and quadrilaterals
- Developing vocabulary to describe special triangles and quadrilaterals
- Generating geometric figures from descriptions of their properties
- Estimating and measuring the size of angles and turns

INVESTIGATION 2

What to Plan Ahead of Time

Materials

- Power Polygons: 1 bucket per 6–8 students (Sessions 1–5 and 8–9)
- Envelopes or resealable plastic bags for storing decks of cards: 1 per student (Sessions 1–3)
- Rubber bands or paper clips for sorted cards: 1 per pair (Sessions 1–3)
- Loop of string that encloses about half the space on the overhead, or a blank transparency (Sessions 1–3)
- Chart paper (Sessions 1–3 and 6–7)
- Overhead projector and pen (Sessions 1–8)
- Blank transparencies (Sessions 6–8)
- Computers (Sessions 4–7 and 9)
- Rulers: 1 per 2–3 students (Session 9)

Other Preparation

- Duplicate student sheets and teaching resources (located at the end of this unit) as follows. If you have Student Activity Booklets, copy only the items marked with an asterisk.

For Sessions 1–3
Guess My Rule Cards (p. 182): 1 deck per pair (preferably on card stock); 1 transparent set*; and 1 set per student (homework)
Student Sheet 6, Is Every Three-Sided Polygon a Triangle? (p. 169): 1 per student (homework)
Student Sheet 7, Is Every Square a Rectangle? Is Every Rectangle a Square? (p. 170): 1 per student (homework)

Student Sheet 8, How to Play Guess My Rule with Shapes (p. 171): 1 per student (homework)

For Sessions 4–5
Student Sheet 3, Coordinate Grids (p. 164): available in class; 2 per student (homework)
Student Sheet 9, Can You Make These Triangles? (p. 172): 1 per student, and 1 transparency*
Student Sheet 10, Can You Make These Quadrilaterals? (p. 173): 1 per student, and 1 transparency*
Student Sheet 11, Find the Fourth Vertex (p. 174): 1 per student (homework)

For Sessions 6–7
Student Sheet 12, Some Shapes Fit Many Categories (p. 175): 1 per student
Student Sheet 13, What Shape Does It Draw? (p. 176): 1 per student (homework)

For Session 8
Student Sheet 14, Angles in the Power Polygons (p. 177): 1 per student
Turtle Turners* (p. 184): 1 transparency per 2 students
Student Sheet 15, Estimating Angles (p. 179): 1 per student

For Session 9
Student Sheet 16, Angles and Turns (p. 180): 1 per student
Student Sheet 17, What Do You Know About 45° and 60° Angles? (p. 181): 1 per student (homework)

Continued on next page

Sessions Within an investigation, the activities are organized by class session, a session being at least a one-hour math class. Sessions are numbered consecutively through an investigation. Often several sessions are grouped together, presenting a block of activities with a single major focus.

When you find a block of sessions presented together—for example, Sessions 1, 2, and 3—read through the entire block first to understand the overall flow and sequence of the activities. Make some preliminary decisions about how you will divide the activities into three sessions for your class, based on what you know about your students. You may need to modify your initial plans as you progress through the activities, and you may want to make notes in the margins of the pages as reminders for the next time you use the unit.

Be sure to read the Session Follow-Up section at the end of the session block to see what homework assignments and extensions are suggested as you make your initial plans.

While you may be used to a curriculum that tells you exactly what each class session should cover, we have found that the teacher is in a better position to make these decisions. Each unit is flexible and may be handled somewhat differently by every teacher. Although we provide guidance for how many sessions a particular group of activities is likely to need, we want you to be active in determining an appropriate pace and the best transition points for your class. It is not unusual for a teacher to spend more or less time than is proposed for the activities.

Ten-Minute Math At the beginning of some sessions, you will find Ten-Minute Math activities. These are designed to be used in tandem with the investigations, but not during the math hour. Rather, we hope you will do them whenever you have a spare 10 minutes—maybe before lunch or recess, or at the end of the day.

Ten-Minute Math offers practice in key concepts, but not always those being covered in the unit. For example, in a unit on using data, Ten-Minute Math might revisit geometric activities done earlier in the year. Complete directions for the suggested activities are included at the end of each unit.

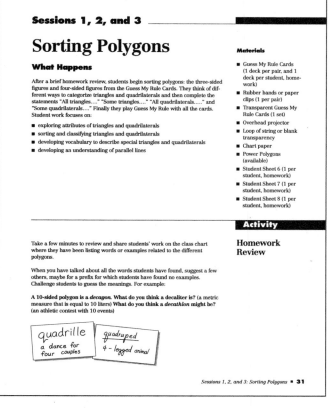

Activities The activities include pair and small-group work, individual tasks, and whole-class discussions. In any case, students are seated together, talking and sharing ideas during all work times. Students most often work cooperatively, although each student may record work individually.

Choice Time In some units, some sessions are structured with activity choices. In these cases, students may work simultaneously on different activities focused on the same mathematical ideas. Students choose which activities they want to do, and they cycle through them.

You will need to decide how to set up and introduce these activities and how to let students make their choices. Some teachers present them as station activities, in different parts of the room. Some list the choices on the board as reminders or have students keep their own lists.

Tips for the Linguistically Diverse Classroom At strategic points in each unit, you will find concrete suggestions for simple modifications of the teaching strategies to encourage the participation of all students. Many of these tips offer alternative ways to elicit critical thinking from students at varying levels of English proficiency, as well as from other students who find it difficult to verbalize their thinking.

The tips are supported by suggestions for specific vocabulary work to help ensure that all students can participate fully in the investigations. The Preview for the Linguistically Diverse Classroom lists important words that are assumed as part of the working vocabulary of the unit. Second-language learners will need to become familiar with these words in order to understand the problems and activities they will be doing. These terms can be incorporated into students' second-language work before or during the unit. Activities that can be used to present the words are found in the appendix, Vocabulary Support for Second-Language Learners. In addition, ideas for making connections to students' languages and cultures, included on the Preview page, help the class explore the unit's concepts from a multicultural perspective.

Session Follow-Up: Homework   In *Investigations,* homework is an extension of classroom work. Sometimes it offers review and practice of work done in class, sometimes preparation for upcoming activities, and sometimes numerical practice that revisits work in earlier units. Homework plays a role both in supporting students' learning and in helping inform families about the ways in which students in this curriculum work with mathematical ideas.

Depending on your school's homework policies and your own judgment, you may want to assign more homework than is suggested in the units. For this purpose you might use the practice pages, included as blackline masters at the end of this unit, to give students additional work with numbers.

For some homework assignments, you will want to adapt the activity to meet the needs of a variety of students in your class: those with special needs, those ready for more challenge, and second-language learners. You might change the numbers in a problem, make the activity more or less complex, or go through a sample activity with

those who need extra help. You can modify any student sheet for either homework or class use. In particular, making numbers in a problem smaller or larger can make the same basic activity appropriate for a wider range of students.

Another issue to consider is how to handle the homework that students bring back to class—how to recognize the work they have done at home without spending too much time on it. Some teachers hold a short group discussion of different approaches to the assignment; others ask students to share and discuss their work with a neighbor; still others post the homework around the room and give students time to tour it briefly. If you want to keep track of homework students bring in, be sure it ends up in a designated place.

Session Follow-Up: Extensions   Sometimes in Session Follow-Up, you will find suggested extension activities. These are opportunities for some or all students to explore a topic in greater depth or in a different context. They are not designed for "fast" students; mathematics is a multifaceted discipline, and different students will want to go further in different investigations. Look for and encourage the sparks of interest and enthusiasm you see in your students, and use the extensions to help them pursue these interests.

Excursions Some of the *Investigations* units include excursions—blocks of activities that could be omitted without harming the integrity of the unit. This is one way of dealing with the great depth and variety of elementary mathematics—much more than a class has time to explore in any one year. Excursions give you the flexibility to make different choices from year to year, doing the excursion in one unit this time, and next year trying another excursion.

Materials

A complete list of the materials needed for teaching this unit follows the unit overview. Some of these materials are available in kits for the *Investigations* curriculum. Individual items can also be purchased from school supply dealers.

Classroom Materials In an active mathematics classroom, certain basic materials should be available at all times: interlocking cubes, pencils, unlined paper, graph paper, calculators, things to count with, and measuring tools. Some activities in this curriculum require scissors and glue sticks or tape. Stick-on notes and large paper are also useful materials throughout.

So that students can independently get what they need at any time, they should know where these materials are kept, how they are stored, and how they are to be returned to the storage area. For example, interlocking cubes are best stored in towers of ten; then, whatever the activity, they should be returned to storage in groups of ten at the end of the hour. You'll find that establishing such routines at the beginning of the year is well worth the time and effort.

Student Sheets and Teaching Resources Student recording sheets and other teaching tools needed for both class and homework are provided as reproducible blackline masters at the end of each unit. We think it's important that students find their own ways of organizing and recording their work. They need to learn how to explain their thinking with both drawings and written words, and how to organize their results so someone else can understand them. For this reason, we deliberately do not provide student sheets for every activity. Regardless of the form in which students do their work, we recommend that they keep their

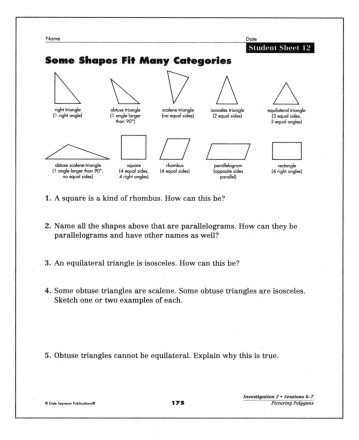

work in a mathematics folder, journal, or notebook so that it is always available to them for reference.

Student Activity Booklets These booklets contain all the sheets each student will need for individual work, freeing you from extensive copying (although you may need or want to copy the occasional teaching resource on transparency film or card stock, or make extra copies of a student sheet).

Calculators and Computers Calculators are used throughout *Investigations*. Many of the units recommend that you have at least one calculator for each pair. You will find calculator activities, plus Teacher Notes discussing this important mathematical tool, in an early unit at each grade level. It is assumed that calculators will be readily available for student use.

Computer activities are offered at all grade levels. How you use the computer activities depends on the number of computers you have available. Technology in the Curriculum discusses ways to incorporate the use of calculators and computers into classroom activities.

Children's Literature   Each unit offers a list of related children's literature that can be used to support the mathematical ideas in the unit. Sometimes an activity is based on a specific children's book, with suggestions for substitutions where practical. While such activities can be adapted and taught without the book, the literature offers a rich introduction and should be used whenever possible.

***Investigations* at Home**   It is a good idea to make your policy on homework explicit to both students and their families when you begin teaching with *Investigations*. How frequently will you be assigning homework? When do you expect homework to be completed and brought back to school? What are your goals in assigning homework? How independent should families expect their children to be? What should the parent's or guardian's role be? The more explicit you can be about your expectations, the better the homework experience will be for everyone.

Investigations at Home (a booklet available separately for each unit, to send home with students) gives you a way to communicate with families about the work students are doing in class. This booklet includes a brief description of every session, a list of the mathematics content emphasized in each investigation, and a discussion of each homework assignment to help families more effectively support their children. Whether or not you are using the *Investigations* at Home booklets, we expect you to make your own choices about homework assignments. Feel free to omit any and to add extra ones you think are appropriate.

Family Letter   A letter that you can send home to students' families is included with the blackline masters for each unit. Families need to be informed about the mathematics work in your classroom; they should be encouraged to participate in and support their children's work. A reminder to send home the letter for each unit appears in one of the early investigations. These letters are also available separately in Spanish, Vietnamese, Cantonese, Hmong, and Cambodian.

Help for You, the Teacher

Because we believe strongly that a new curriculum must help teachers think in new ways about mathematics and about their students' mathematical thinking processes, we have included a great deal of material to help you learn more about both.

About the Mathematics in This Unit   This introductory section summarizes the critical information about the mathematics you will be teaching. It describes the unit's central mathematical ideas and the ways students will encounter them through the unit's activities.

About the Assessment in this Unit   This introductory section highlights Teacher Checkpoints and assessment activities contained in the unit. It offers questions to stimulate your assessment as you observe the development of students' mathematical thinking and learning.

Teacher Notes   These reference notes provide practical information about the mathematics you are teaching and about our experience with how students learn. Many of the notes were written in response to actual questions from teachers or to discuss important things we saw happening in the

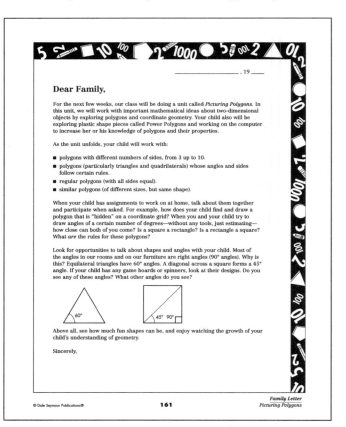

field-test classrooms. Some teachers like to read them all before starting the unit, then review them as they come up in particular investigations.

Dialogue Boxes   Sample dialogues demonstrate how students typically express their mathematical ideas, what issues and confusions arise in their thinking, and how some teachers have guided class discussions.

These dialogues are based on the extensive classroom testing of this curriculum; many are word-for-word transcriptions of recorded class discussions. They are not always easy reading; sometimes it may take some effort to unravel what the students are trying to say. But this is the value of these dialogues; they offer good clues to how your students may develop and express their approaches and strategies, helping you prepare for your own class discussions.

Where to Start   You may not have time to read everything the first time you use this unit. As a first-time user, you will likely focus on understanding the activities and working them out with your students. Read completely through all the activities before starting to present them. Also read those sections listed in the Contents under the heading Where to Start.

Teacher Note ▷ *Beyond Vocabulary*

The work in this unit will bring up many vocabulary words that students do not yet know. Avoid the temptation to stop to teach a lesson specifically on vocabulary or to test students on vocabulary words. This temptation is especially great in the beginning of Investigation 2, when students are introduced to a large amount of vocabulary as they try to describe polygons by the number of sides, and to describe kinds and characteristics of triangles and quadrilaterals.

Throughout the unit, you and your students will be using words that help you describe different kinds of polygons and angles. While students are working together with polygons on and off the computer, they will talk about the shapes in their own words. For example, a "pointy angle" may mean an acute angle and a "wide angle" an obtuse one. However, the activities in this unit will motivate students to use more precise language. As students work to make themselves clear, you will hear some unusual, but accurate, descriptions. Try to achieve a balance between using the most accurate words and agreeing on classroom vocabulary that makes the most sense to students.

Students will learn the mathematical terms as they hear them in context. Use the correct terms yourself and informally explain to students what they mean. As students hear you and others use the terms, they will adopt the ones that make sense to them—but this process will take time. Do not expect students to use terms as soon as they hear them. The activities in this unit provide repeated experience with the same terms and concepts, so students will have a number of opportunities to pick up the new vocabulary. Be aware that some students will adopt it easily, while others will need more time and exposure.

More important is that students develop accurate concepts for shapes for which they already have names. Many 10-year-olds (and even older students) have very limited ideas about common shapes. For example, they may call a rectangle a square; they believe that a triangle must have symmetry (must have at least two equal sides) and must not be tilted on the page. Help students build concepts without giving them definitions. Allow them to try out their ideas in class discussions, but expect them to make themselves clear and to justify their conjectures.

DIALOGUE BOX

Are All Three-Sided Polygons Triangles?

As this class discusses categories for the triangles on the Guess My Rule Cards, some students are confused about whether *all* three-sided polygons should be called triangles.

We're looking for categories for sorting three-sided polygons. So far we have shapes with two equal sides, shapes with all equal sides, shapes with all unequal sides, and shapes that if you had another shape just like it, you could make a square or rectangle. What else?

Robby: How about shapes that look like triangles. Examples are 3, 11, 6, 8, and 12. Not 9.

What must a figure have to look like a triangle?

Robby: Three equal sides. Like one would be left side 15 cm, right side 15 cm, and the bottom connecting the two 15 cm.

Robby feels the other three-sided polygons aren't triangles. What do the rest of you think? Does a polygon have to have three equal sides to be called a triangle?

Yu-Wei: The shape of 9 is a stretched triangle, so it's not a triangle.

Julie: I disagree, because other ones that are sort of stretched like 6 and 8 *are* triangles.

Robby, which ones do you feel are not triangles?

Robby: I think 7, 10, 9, 2, and 5 aren't.

[There is a lot of discussion at this point, and the teacher asks to hear more about what makes a shape a triangle.]

Help! What *is* a definition of a triangle?

Lindsay: It has to have three angles. You know that because *tri*-angle; *tri* means "three."

If *triangle* means three angles, where do the sides fit in?

Matt: Anything with three sides is a triangle.

Manuel: I agree with three sides, but I think they all have to be equal.

Leon: Then how come they have triangles called isosceles and scalene?

[Class ends with the question unresolved. Discussion continues the next day.]

We ended math yesterday with a big question: Are all three-sided polygons triangles? What did you decide?

Cara: I wrote: "I think a triangle is a shape that has three sides and three corners. I think that triangles don't have to have the same length in the sides."

Robby: I made a category called "Triangles," and I wrote, "A triangle should have even sides and three corners."

So what about three-sided shapes that don't have even sides? What would you call them?

Robby: I don't know.

Antonio: I used to think like Robby, but now I think any three-sided polygon is a triangle. I didn't know what to call those other shapes except triangles.

Julie: I agree. I think a triangle's something with three sides no matter how long they are.

Manuel: Triangles can have different sides, and then have other names that mean they're not equal, like *right* or *isosceles* triangles.

It's important that we're coming to agreement about these terms as a class.

The *Investigations* curriculum incorporates the use of two forms of technology in the classroom: calculators and computers. Calculators are assumed to be standard classroom materials, available for student use in any unit. Computers are explicitly linked to one or more units at each grade level; they are used with the unit on 2-D geometry at each grade, as well as with some of the units on measuring, data, and changes.

Using Calculators

In this curriculum, calculators are considered tools for doing mathematics, similar to pattern blocks or interlocking cubes. Just as with other tools, students must learn both *how* to use calculators correctly and *when* they are appropriate to use. This knowledge is crucial for daily life, as calculators are now a standard way of handling numerical operations, both at work and at home.

Using a calculator correctly is not a simple task; it depends on a good knowledge of the four operations and of the number system, so that students can select suitable calculations and also determine what a reasonable result would be. These skills are the basis of any work with numbers, whether or not a calculator is involved.

Unfortunately, calculators are often seen as tools to check computations with, as if other methods are somehow more fallible. Students need to understand that any computational method can be used to check any other; it's just as easy to make a mistake on the calculator as it is to make a mistake on paper or with mental arithmetic. Throughout this curriculum, we encourage students to solve computation problems in more than one way in order to double-check their accuracy. We present mental arithmetic, paper-and-pencil computation, and calculators as three possible approaches.

In this curriculum we also recognize that, despite their importance, calculators are not always appropriate in mathematics instruction. Like any tools, calculators are useful for some tasks but not for others. You will need to make decisions about when to allow students access to calculators and when to ask that they solve problems without them so that they can concentrate on other tools and skills. At times when calculators are or are not appropriate for a particular activity, we make specific recommendations. Help your students develop their own sense of which problems they can tackle with their own reasoning and which ones might be better solved with a combination of their own reasoning and the calculator.

Managing calculators in your classroom so that they are a tool, and not a distraction, requires some planning. When calculators are first introduced, students often want to use them for everything, even problems that can be solved quite simply by other methods. However, once the novelty wears off, students are just as interested in developing their own strategies, especially when these strategies are emphasized and valued in the classroom. Over time, students will come to recognize the ease and value of solving problems mentally, with paper and pencil, or with manipulatives, while also understanding the power of the calculator to facilitate work with larger numbers.

Experience shows that if calculators are available only occasionally, students become excited and distracted when they are permitted to use them. They focus on the tool rather than on the mathematics. In order to learn when calculators are appropriate and when they are not, students must have easy access to them and use them routinely in their work.

If you have a calculator for each student, and if you think your students can accept the responsibility, you might allow them to keep their calculators with the rest of their individual materials, at least for the first few weeks of school. Alternatively, you might store them in boxes on a shelf, number each calculator, and assign a corresponding number to each student. This system can give students a sense of ownership while also helping you keep track of the calculators.

Using Computers

Students can use computers to approach and visualize mathematical situations in new ways. The computer allows students to construct and manipulate geometric shapes, see objects move according to rules they specify, and turn, flip, and repeat a pattern.

This curriculum calls for computers in units where they are a particularly effective tool for learning mathematics content. One unit on 2-D geometry at each of the grades 3–5 includes a core of activities that rely on access to computers, either in the classroom or in a lab. Other units on geometry, measuring, data, and changes include computer activities, but can be taught without them. In these units, however, students' experience is greatly enhanced by computer use.

The following list outlines the recommended use of computers in this curriculum:

Kindergarten
Unit: *Making Shapes and Building Blocks*
 (Exploring Geometry)
Software: *Shapes*
Source: provided with the unit

Grade 1
Unit: *Survey Questions and Secret Rules*
 (Collecting and Sorting Data)
Software: *Tabletop, Jr.*
Source: Broderbund

Unit: *Quilt Squares and Block Towns*
 (2-D and 3-D Geometry)
Software: *Shapes*
Source: provided with the unit

Grade 2
Unit: *Mathematical Thinking at Grade 2*
 (Introduction)
Software: *Shapes*
Source: provided with the unit

Unit: *Shapes, Halves, and Symmetry*
 (Geometry and Fractions)
Software: *Shapes*
Source: provided with the unit

Unit: *How Long? How Far?* (Measuring)
Software: *Geo-Logo*
Source: provided with the unit

Grade 3
Unit: *Flips, Turns, and Area* (2-D Geometry)
Software: *Tumbling Tetrominoes*
Source: provided with the unit

Unit: *Turtle Paths* (2-D Geometry)
Software: *Geo-Logo*
Source: provided with the unit

Grade 4
Unit: *Sunken Ships and Grid Patterns*
 (2-D Geometry)
Software: *Geo-Logo*
Source: provided with the unit

Grade 5
Unit: *Picturing Polygons* (2-D Geometry)
Software: *Geo-Logo*
Source: provided with the unit

Unit: *Patterns of Change* (Tables and Graphs)
Software: *Trips*
Source: provided with the unit

Unit: *Data: Kids, Cats, and Ads* (Statistics)
Software: *Tabletop, Sr.*
Source: Broderbund

The software provided with the *Investigations* units uses the power of the computer to help students explore mathematical ideas and relationships that cannot be explored in the same way with physical materials. With the *Shapes* (grades 1–2) and *Tumbling Tetrominoes* (grade 3) software, students explore symmetry, pattern, rotation and reflection, area, and characteristics of 2-D shapes. With the *Geo-Logo* software (grades 2–5), students investigate rotations and reflections, coordinate geometry, the properties of 2-D shapes, and angles. The *Trips* software (grade 5) is a mathematical exploration of motion in which students run experiments and interpret data presented in graphs and tables.

We suggest that students work in pairs on the computer; this not only maximizes computer resources but also encourages students to consult, monitor, and teach each other. Generally, more than two students at one computer find it difficult to share. Managing access to computers is an issue for every classroom. The curriculum gives you explicit support for setting up a system. The units are structured on the assumption that you have enough computers for half your students to work on the machines in pairs at one time. If you do not have access to that many computers, suggestions are made for structuring class time to use the unit with fewer than five.

Assessment plays a critical role in teaching and learning, and it is an integral part of the *Investigations* curriculum. For a teacher using these units, assessment is an ongoing process. You observe students' discussions and explanations of their strategies on a daily basis and examine their work as it evolves. While students are busy recording and representing their work, working on projects, sharing with partners, and playing mathematical games, you have many opportunities to observe their mathematical thinking. What you learn through observation guides your decisions about how to proceed. In any of the units, you will repeatedly consider questions like these:

■ Do students come up with their own strategies for solving problems, or do they expect others to tell them what to do? What do their strategies reveal about their mathematical understanding?

■ Do students understand that there are different strategies for solving problems? Do they articulate their strategies and try to understand other students' strategies?

■ How effectively do students use materials as tools to help with their mathematical work?

■ Do students have effective ideas for keeping track of and recording their work? Do keeping track of and recording their work seem difficult for them?

You will need to develop a comfortable and efficient system for recording and keeping track of your observations. Some teachers keep a clipboard handy and jot notes on a class list or on adhesive labels that are later transferred to student files. Others keep loose-leaf notebooks with a page for each student and make weekly notes about what they have observed in class.

Assessment Tools in the Unit

With the activities in each unit, you will find questions to guide your thinking while observing the students at work. You will also find two built-in assessment tools: Teacher Checkpoints and embedded Assessment activities.

Teacher Checkpoints   The designated Teacher Checkpoints in each unit offer a time to "check in" with individual students, watch them at work, and ask questions that illuminate how they are thinking.

At first it may be hard to know what to look for, hard to know what kinds of questions to ask. Students may be reluctant to talk; they may not be accustomed to having the teacher ask them about their work, or they may not know how to explain their thinking. Two important ingredients of this process are asking students open-ended questions about their work and showing genuine interest in how they are approaching the task. When students see that you are interested in their thinking and are counting on them to come up with their own ways of solving problems, they may surprise you with the depth of their understanding.

Teacher Checkpoints also give you the chance to pause in the teaching sequence and reflect on how your class is doing overall. Think about whether you need to adjust your pacing: Are most students fluent with strategies for solving a particular kind of problem? Are they just starting to formulate good strategies? Or are they still struggling with how to start? Depending on what you see as the students work, you may want to spend more time on similar problems, change some of the problems to use smaller numbers, move quickly to more-challenging material, modify subsequent activities for some students, work on particular ideas with a small group, or pair students who have good strategies with those who are having more difficulty.

Embedded Assessment Activities   Assessment activities embedded in each unit will help you examine specific pieces of student work, figure out what they mean, and provide feedback. From the students' point of view, these assessment activities are no different from any others. Each is a learning experience in and of itself, as well as an opportunity for you to gather evidence about students' mathematical understanding.

The embedded assessment activities sometimes involve writing and reflecting; at other times, a discussion or brief interaction between student and teacher; and in still other instances, the creation and explanation of a product. In most cases, the assessments require that students *show* what they did, *write* or *talk* about it, or do both. Having to explain how they worked through a problem helps students be more focused and clear in their mathematical thinking. It also helps them realize that doing mathematics is a process that may involve tentative starts, revising one's approach, taking different paths, and working through ideas.

Teachers often find the hardest part of assessment to be interpreting their students' work. We provide guidelines to help with that interpretation. If you have used a process approach to teaching writing, the assessment in *Investigations* will seem familiar. For many of the assessment activities, a Teacher Note provides examples of student work and a commentary on what it indicates about student thinking.

Documentation of Student Growth

To form an overall picture of mathematical progress, it is important to document each student's work. Many teachers have students keep their work in folders, notebooks, or journals, and some like to have students summarize their learning in journals at the end of each unit. It's important to document students' progress, and we recommend that you keep a portfolio of selected work for each student, unit by unit, for the entire year. The final activity in each *Investigations* unit, called Choosing Student Work to Save, helps you and the students select representative samples for a record of their work.

This kind of regular documentation helps you synthesize information about each student as a mathematical learner. From different pieces of evidence, you can put together the big picture. This synthesis will be invaluable in thinking about where to go next with a particular child, deciding where more work is needed, or explaining to parents (or other teachers) how a child is doing.

If you use portfolios, you need to collect a good balance of work, yet avoid being swamped with an overwhelming amount of paper. Following are some tips for effective portfolios:

- Collect a representative sample of work, including some pieces that students themselves select for inclusion in the portfolio. There should be just a few pieces for each unit, showing different kinds of work—some assignments that involve writing as well as some that do not.

- If students do not date their work, do so yourself so that you can reconstruct the order in which pieces were done.

- Include your reflections on the work. When you are looking back over the whole year, such comments are reminders of what seemed especially interesting about a particular piece; they can also be helpful to other teachers and to parents. Older students should be encouraged to write their own reflections about their work.

Assessment Overview

There are two places to turn for a preview of the assessment opportunities in each *Investigations* unit. The Assessment Resources column in the unit Overview Chart identifies the Teacher Checkpoints and Assessment activities embedded in each investigation, guidelines for observing the students that appear within classroom activities, and any Teacher Notes and Dialogue Boxes that explain what to look for and what types of student responses you might expect to see in your classroom. Additionally, the section About the Assessment in This Unit gives you a detailed list of questions for each investigation, keyed to the mathematical emphases, to help you observe student growth.

Depending on your situation, you may want to provide additional assessment opportunities. Most of the investigations lend themselves to more frequent assessment, simply by having students do more writing and recording while they are working.

Picturing Polygons

Content of This Unit   As students create polygons with shape pieces, they construct, apply, discuss, and evaluate mathematical definitions of these shapes. Analyzing the properties of polygons helps students draw them on coordinate grids, both on and off the computer. Students closely investigate various properties of triangles, quadrilaterals, and regular polygons, asking which remain constant and which change when larger and smaller similar shapes are made. Using *Geo-Logo's* turtle commands, they write procedures to draw regular polygons; they also measure side lengths and angles of polygons, and look at patterns in sums of angles and of turns.

Connections with Other Units   If you are doing the full-year *Investigations* curriculum in the suggested sequence for grade 5, this is the second of nine units. If your students have used the 2-D Geometry unit for grade 3, *Turtle Paths,* or grade 4, *Sunken Ships and Grid Patterns,* they have worked with *Geo-Logo.* If they have not had experience with either *Logo* or *Geo-Logo,* you may need to take more time with the computer activities in this unit.

In this unit, students find patterns in the relationship between side lengths and area as they make similar shapes. Students will continue to find patterns in growing geometric designs in the grade 5 Tables and Graphs unit, *Patterns of Change.*

If your school is not using the full-year curriculum, this unit can also be used successfully at grade 6.

Investigations Curriculum ■ Suggested Grade 5 Sequence

Mathematical Thinking at Grade 5 (Introduction and Landmarks in the Number System)

▶ *Picturing Polygons* (2-D Geometry)

Name That Portion (Fractions, Percents, and Decimals)

Between Never and Always (Probability)

Building on Numbers You Know (Computation and Estimation Strategies)

Measurement Benchmarks (Estimating and Measuring)

Patterns of Change (Tables and Graphs)

Containers and Cubes (3-D Geometry: Volume)

Data: Kids, Cats, and Ads (Statistics)

Investigation 1 ▪ Identifying Polygons

Class Sessions	Activities	Pacing
Session 1 (p. 5) IS IT A POLYGON?	Picking Out Polygons Defining a Polygon Polygon Art Homework: Picasso's Polygons	minimum 1 hr
Session 2 (p. 9) MAKING POLYGONS	Putting Together Power Polygons Polygon Pictures Homework: Coordinate Grids	minimum 1 hr
Session 3 (p. 15) POLYGON PICTURES WITH COORDINATE GEOMETRY	Finding the Coordinates Teacher Checkpoint: Identifying Coordinates Homework: Hidden Polygon Pictures	minimum 1 hr
Session 4 (p. 19) COORDINATE GEOMETRY WITH *GEO-LOGO*	Hidden Picture Homework Writing *Geo-Logo* Commands Off Computer: Brainstorming the Names of Polygons On Computer: Making Polygon Pictures with *Geo-Logo* Pooling Polygon Names Homework: Types of Polygons	minimum 1 hr

◗ **Ten-Minute Math** ▪ **Multiple Bingo**

Mathematical Emphasis

- Distinguishing between polygons and shapes that are not polygons

- Drawing polygons

- Locating points on a coordinate grid

- Using the *Geo-Logo* commands `setxy` and `jumpto` to draw polygons on the computer

- Recognizing and naming polygons by number of sides

Assessment Resources

Describing Polygons (Dialogue Box, p. 8)

Teacher Checkpoint: Identifying Coordinates (p. 17)

Putting It All Together (Teacher Note, p. 18)

Materials

Overhead projector and pen

Blank transparency

Chart paper

Power Polygons

Unlined paper

Colored pencils, crayons, or markers

Computers

Apple Macintosh disk, *Geo-Logo*™ for *Picturing Polygons*

Large-screen monitor

Student Sheets 1–5

Teaching resource sheets

Family letter

Investigation 2 ■ Triangles and Quadrilaterals

Class Sessions	Activities	Pacing

Sessions 1, 2, and 3 (p. 31)
SORTING POLYGONS
	Homework Review
	Identifying Triangles by Their Angles
	Sorting Triangles
	All Triangles, Some Triangles
	Identifying Quadrilaterals
	Sorting Quadrilaterals
	Attributes of Quadrilaterals
	Playing Guess My Rule
	Homework: Is Every Three-Sided Polygon a Triangle?;
	   Is Every Square a Rectangle? Is Every Rectangle a
	   Square?; How to Play Guess My Rule with Shapes
		minimum
		3 hr

Sessions 4 and 5 (p. 46) MAKING SHAPES THAT FOLLOW RULES	Adding More Attributes Following the Rules Can You Make These? On Computer: Making Shapes with *Geo-Logo* Off Computer: Making Shapes with Power Polygons Discussing the Shapes Homework: Coordinate Grids and "Quad" or "Tri" Commands; Find the Fourth Vertex Extension: Writing *Geo-Logo* Procedures	minimum 2 hr
Sessions 6 and 7 (p. 56) USING MOVE AND TURN COMMANDS	Learning the Move and Turn Commands Learning the Repeat Command On Computer: Polygons with Moves and Turns Assessment: Shapes That Fit Many Categories How Did You Make the Equilateral Triangle? Comparing Turns and Angles Homework: What Shape Does It Draw?	minimum 2 hr
Session 8 (p. 70) FINDING ANGLE SIZES	Finding Angle Sizes Teacher Checkpoint: Angles in the Power Polygons Discussion: How We Found the Angles Homework: Estimating Angles	minimum 1 hr
Session 9 (p. 76) ANGLES AND TURNS TOGETHER	On Computer: Angle and Turn Game Off Computer: Measuring Turns and Angles Off Computer: Estimating Angle Size Writing About Angles Homework: What Do You Know About 45° and 60° Angles?	minimum 1 hr

◓ Ten-Minute Math ■ Factor Bingo

Mathematical Emphasis

- Reasoning and communicating about properties of geometric shapes
- Sorting and classifying triangles and quadrilaterals
- Developing vocabulary to describe special triangles and quadrilaterals
- Generating geometric figures from descriptions of their properties
- Estimating and measuring the size of angles and turns

Assessment Resources

Dialogue Boxes: Are All Three-Sided Polygons Triangles? (p. 44); Are Squares Rectangles? (p. 45); What Rules Are You Following? (p. 54); Which Are Impossible? (p. 55); Finding Angle Measures of Power Polygons (p. 74)

Observing the Students (p. 61)

Assessment: Shapes That Fit Many Categories (p. 61)

Teacher Notes: Assessment: Shapes That Fit Many Categories (p. 66); The Rule of 180° (p. 68)

Teacher Checkpoint: Angles in the Power Polygons (p. 72)

Materials

Power Polygons

Envelopes or resealable plastic bags

Rubber bands or paper clips

String

Chart paper

Overhead projector

Blank transparencies

Computers, large-screen monitor

Rulers

Student Sheets 6–17

Teaching resource sheets

Investigation 3 ▪ Regular Polygons and Similarity

Class Sessions	Activities	Pacing
Sessions 1 and 2 (p. 83) REGULAR POLYGONS	Defining Regular Polygons On Computer: Guess and Test Polygons Finding Total Turns and Angles On Computer: Drawing Regular Polygons Off Computer: Turns and Angles Homework: Total Turns and Angles Extension: Writing Procedures with Angles	minimum 2 hr
Session 3 (p. 90) PATTERNS OF ANGLES AND TURNS	Discussion: Patterns for Regular Polygons Assessment: Which Are Regular Polygons? Homework: Polygons That Are Not Regular	minimum 1 hr
Session 4 (p. 93) BUILDING SIMILAR SHAPES	Pooling Homework Results Exploring Similarity Building Similar Shapes Generalizing About Growing Shapes Homework: Length of Sides Versus Area	minimum 1 hr
Sessions 5 and 6 (p. 100) SIMILARITY ACTIVITIES	On Computer: Similar Rectangles On Computer: Similar Houses Off Computer: Similarity Poster Sharing Conclusions About Similar Shapes Choosing Student Work to Save Homework: Drawing More Similar Rectangles Extension: Growth of Perimeter and Area in Similar Shapes	minimum 2 hr

Mathematical Emphasis

- Distinguishing between regular and nonregular polygons

- Exploring the relationship between the number of sides a polygon has and the sums of its turns and angles

- Exploring relationships among angles, line lengths, and areas of similar polygons

- Comparing areas of shapes

Assessment Resources

Drawing Regular Polygons (Dialogue Box, p. 89)

Assessment: Which Are Regular Polygons? (p. 91)

Choosing Student Work to Save (p. 107)

Writing Procedures for Similar Figures (Teacher Note, p. 109)

Building Larger Similar Figures (Teacher Note, p. 110)

Materials

Power Polygons

Computers

Calculators

Students' transparent Turtle Turners

Rulers

Overhead projector

Large paper

Stick-on notes

Crayons, colored markers, or pencils

Student Sheets 18–25

Teaching resource sheets

Following are the basic materials needed for the activities in this unit. Many of the items can be purchased from the publisher, either individually or in the Teacher Resource Package and the Student Materials Kit for grade 5. Detailed information is available on the *Investigations* order form. To obtain this form, call toll-free 1-800-872-1100 and ask for a Dale Seymour customer service representative.

Power Polygons: 1 bucket per 6–8 students

Note: Power Polygons include the six basic pattern-block shapes plus 9 more related shapes. Many of the activities in this unit are based on the specific pieces in the Power Polygons set. For those activities, substitutions of shape pieces without the same size relationships will not work.

Apple Macintosh disk, *Geo-Logo* for *Picturing Polygons* (packaged with this book)

Computers—Macintosh II or above, with 4 MB of internal memory (RAM) and Apple System Software 7.0 or later. Optimum: 1 for every 2 students. Minimum: 1 for every 4–6 students. With fewer computers, you will need to modify how you manage the unit; see Managing the Computer Activities in This Unit, p. I-21.

Projection device or large-screen monitor on one computer for whole-class viewing (recommended)

Calculators: 1 per small group

Large paper for posters (about 11" by 17" inches): 1 per student

Envelopes or resealable plastic bags for storing decks of cards: 1 per student

Rubber bands or paper clips: 1 per pair

Rulers: 1 per student

Stick-on notes: 1 pad

String loops (for Guess My Rule)

Colored pencils, markers, or crayons

Chart paper

Unlined paper

Overhead projector

Blank overhead transparencies and pens

The following materials are provided at the end of this unit as blackline masters. A Student Activity Booklet containing all student sheets and teaching resources needed for individual work is available.

Family Letter (p. 161)

Student Sheets 1–25 (p. 162)

Teaching Resources:

> Polygons and Other Figures (p. 167)
>
> *Geo-Logo* User Sheet (p. 168)
>
> Guess My Rule Cards (p. 182)
>
> Turtle Turners (p. 184)
>
> Polygons: Regular and Not Regular (p. 194)
>
> One-Centimeter Graph Paper (p. 195)
>
> Multiple Bingo: 100 Chart (p. 196)
>
> Multiple Bingo: 300 Chart (p. 197)
>
> Multiple Bingo Cards (p. 198)
>
> Factor Bingo: Multiplication Table (p. 199)
>
> Factor Bingo Cards (p. 200)

Practice Pages (p. 201)

Related Children's Literature

Froman, Robert. *Angles Are Easy as Pie*. New York: Thomas Y. Crowell, 1975.

Many people think of geometry in only two ways: as learning simple shapes and their names, and as doing complicated proofs. The geometry in between these two extremes, however, is rich with mathematical possibilities. In this unit, students investigate two-dimensional shapes in depth. They describe, draw, combine, and define polygons. As students build polygons by combining shapes, they build visualization skills. As they describe and sort polygons, they learn about properties of polygons (such as angle measure and parallelism), and they reason and communicate about mathematical ideas.

One important mathematical idea students will encounter in this unit is that many definitions are possible. As students describe the characteristics or attributes of polygons, they construct the idea that mathematics is not just "knowing" the definition. Real mathematical activity includes forming, applying, discussing, and evaluating definitions. In this unit, students define polygons by describing their properties, by specifying the coordinates of their corners (vertices) on coordinate grids on and off the computer, and by writing forward and turn computer commands that draw polygons.

Students experience number and geometry as interrelated. One of the ways mathematicians connect number and geometry is through coordinate graphing—specifying positions in space with numbers. Understanding coordinates is important for reading and constructing graphs and maps and for graphing functions. Students connect number and geometry when they make shapes on the computer using coordinate commands and also when they use motions and turns. When students write commands that instruct an on-screen "turtle" to draw polygons, they specify numerical measurements (lengths, turns, and angles) explicitly. This type of drawing—compared with drawing by hand—requires that students translate what they *see and understand intuitively* into what they can *state explicitly.*

In a similar vein, students explore connections between geometry and number by investigating similar figures. They use manipulatives and the computer to investigate connections between polygons that have the same shape but different sizes. They look for patterns in the growth of the side lengths and the growth of the areas of the shapes. Students come to this with an expectation that shapes grow by adding to the sides, but find through trial and error on the computer that simply adding the same amount to each side of a polygon does not necessarily produce a similar shape. They experience the proportional relationship between shapes that are similar—sides of similar shapes are *multiples* of one another. Students also discover that computers can be a useful tool for mathematical exploration and thinking.

At the beginning of each investigation, the Mathematical Emphasis section tells you what is most important for students to learn about during that investigation. Students are not expected to "master" all of these understandings and processes. Rather, they are gradually learning more and more about each idea over many years of schooling. Individual students will begin and end the unit with different levels of knowledge and skill, but all will gain greater knowledge about two-dimensional space and shape and develop strategies for solving problems involving these ideas.

Throughout the *Investigations* curriculum, there are many opportunities for ongoing daily assessment as you observe, listen to, and interact with students at work. In this unit, you will find two Teacher Checkpoints:

Investigation 1, Session 3:
Identifying Coordinates (p. 17)

Investigation 2, Session 8:
Angles in the Power Polygons (p. 72)

This unit also has two embedded assessment activities:

Investigation 2, Sessions 6–7:
Shapes That Fit Many Categories (p. 61)

Investigation 3, Session 3:
Which Are Regular Polygons? (p. 91)

In addition, you can use almost any activity in this unit to assess your students' needs and strengths. Listed below are questions to help you focus your observation in each investigation. You may want to keep track of your observations for each student to help you plan your curriculum and monitor students' growth. Suggestions for documenting student growth can be found in the section About Assessment.

Investigation 1: Identifying Polygons

- How do students describe a polygon? What information do students use when identifying polygons?

- How do students approach the task of drawing polygons? How do they use power polygons to make three-sided shapes? Four-sided shapes? Shapes with more than four and up to ten sides?

- How do students describe points on a coordinate grid? How do they interpret coordinates? Do they go over 10 and up 30 from the origin? Do they know how to use *Geo-Logo* commands to draw polygons?

- How do students recognize and name polygons? Are they accurate when they count the sides? As they are counting, how do they keep track of the sides they've counted?

- How do students refer to various polygons? Do they know the names of the shapes? Do they use mathematical terminology?

Investigation 2: Triangles and Quadrilaterals

- What terminology do students use when referring to various shapes and their properties? What information do they add to class charts of polygons? Are they bound by images of "traditional" shapes they may have seen, or are they flexible in identifying triangles and quadrilaterals?

- How do students sort and classify polygons and quadrilaterals when playing Guess My Rule? How do they justify whether a shape fits or does not fit into a category? How many different rules can they come up with for each session of Guess My Rule?

- How do students refer to "special" triangles and quadrilaterals? What language do they use to describe these shapes? How do students identify characteristics of these shapes?

- Do students' polygons fit a specified set of rules they've been asked to follow? How do they use these rules to create polygons on the computer? How do they justify their polygons?

- Do students understand the difference between turns and angles? How do students find the measure of unknown angles? How do students use a straight line (180°) or a circle (360°) as a reference point for estimating angles? What strategies do students use when asked to reproduce certain angles?

Investigation 3: Regular Polygons and Similarity

- How do students define regular and nonregular polygons? How do they distinguish between them?

- How do students use what they know about the number of sides and the sum of the angles in a shape to predict what the sum of the angles in another shape will be? Do they notice the relationship between the number of sides a polygon has and the sum of its turns and angles? What patterns do students notice as the number of sides increases in a regular polygon?

- What do students notice about the relationship between line lengths and area? What patterns do they use to figure out how many of an original shape fit into a larger but similar shape? Do students use a consistent strategy when finding the relationship between similar figures?

- How do students determine the area of a shape as it increases proportionally to form larger but similar shapes? How do students show their understanding of the relationship of area to perimeter in similar figures?

In the *Investigations* curriculum, mathematical vocabulary is introduced naturally during the activities. We don't ask students to learn definitions of new terms; rather, they come to understand such words as *factor* or *area* or *symmetry* by hearing them used frequently in discussion as they investigate new concepts. This approach is compatible with current theories of second-language acquisition, which emphasize the use of new vocabulary in meaningful contexts while students are actively involved with objects, pictures, and physical movement.

Listed below are some key words used in this unit that will not be new to most English speakers at this age level, but may be unfamiliar to students with limited English proficiency. You will want to spend additional time working on these words with your students who are learning English. If your students are working with a second-language teacher, you might enlist your colleague's aid in familiarizing students with these words, before and during this unit. In the classroom, look for opportunities for students to hear and use these words. Activities you can use to present the words are given in the appendix, Vocabulary Support for Second-Language Learners (p. 114).

shape, sides, triangle, quadrilateral, impossible Students work with the basic vocabulary of 2-D geometry as they describe the *shapes* and the number of *sides* of polygons, starting with *triangles* and *quadrilaterals*. As students construct polygons, they learn that some shapes are *impossible* to make, for example, a triangle with parallel sides.

doubled, tripled, original When students learn about similar polygons, they use computer software that allows them to draw a polygon and then stretch it so that the sides of the *original* figure are *doubled* and *tripled*.

Multicultural Extensions for All Students

Whenever possible, encourage students to share words, objects, customs, or any aspects of daily life from their own cultures and backgrounds that are relevant to the activities in this unit.

For example, in English, the names of polygons are formed with prefixes that denote the number of sides (*tri*angle, *three* sides); the Teacher Note, Types of Polygons (p. 27), contains a list of these specific names. During activities on naming polygons, students fluent in other languages can explore whether the names of polygons in their languages are similarly formed and report their findings to the class.

Students may also be able to share information about landmarks in other countries that have distinctive polygonal shapes, such as the face of a building or monument, bringing in pictures if possible.

The class might be interested in exploring the language of signs to see whether certain polygonal shapes that we recognize as having a particular meaning carry the same meaning internationally; for example, the octagonal sign for STOP.

This unit is dependent on having students use computers on an ongoing basis. Ideally, students working in pairs will use computers almost daily for approximately 20 minutes or more. You will need to plan your sessions so that all students have ample time to do the computer activities.

The structure of this unit supposes enough computers in the classroom so that about half the students can work on computers, in pairs, at one time. Having five to eight computers available all day in a classroom is most conducive to effective and efficient use of these resources. At present, you may not have access to this many computers, but here is a model for the way computers may be integrated in classrooms in the near future. Below are suggestions for modifying the management of computer resources where there are fewer than five computers available and where there is a computer laboratory available.

Structuring the Unit to Match Computer Availability

Five to Eight Computers Five to eight computers in the room is enough so that half of the students, working in pairs, can use them while the other half work on an off-computer activity. The two groups then switch halfway though the work time. Continuing to allow pairs of students to work at the computers at other times during the school day will give them the chance to complete all the computer activities, plus time to repeat some activities using different commands or strategies.

Computer Laboratory If you have a computer laboratory, you may want to involve the whole class in computer activities at the same time. Specific suggestions for modifying the sessions in this way are found in the overview for each investigation (pp. 4, 30, and 81).

Fewer than Five Computers If you have fewer than five computers available, pairs will need to use the computers throughout the school day, in rotation, so that everyone has completed the activities before the follow-up discussion. With this strategy, you will always introduce the on-computer activities before the off-computer activities. Students can begin cycling through the computer activity as you work with the remainder of the class on the off-computer activities.

Working at the Computer

Working in Pairs Students should work on the computers in pairs (or threes). This approach not only maximizes computer resources but also encourages students to consult, monitor, and teach each other. Generally, more than two students at one computer find it difficult to share. Since students will often not complete their computer work in one session, plan that they stay with the same partner for the entire unit.

Saving Student Work Students will want to save their work on the computer in some of the activities. This can be done in two ways: (1) They can use the same computer each time and save their work on the computer's internal drive, or (2) they can save their work on a disk that they can use on any of the available computers. If you can provide pairs of students with their own disks, computer management will be simpler, because each pair will be able to use any computer when it is available. If students' work is saved on a particular computer, they may have to wait until other students using that computer are finished. Instructions for saving work are in the *Geo-Logo* Teacher Tutorial (p. 115).

Demonstrating Computer Activities Often you will need to use a computer with the whole class to demonstrate computer activities and to share results during whole-group discussions. It is helpful if a computer is connected to a large-screen monitor or projection device—a "large display." If no large display is available, gather groups of students as close as possible around the computer. Increasing the font size when entering commands will make the commands more visible. To increase the font size, choose **All Large** under the **Font** menu. When you are finished demonstrating, return the font to its regular size by choosing **All Small** under the **Font** menu.

If your computer display is very small and it is difficult for students to see the demonstrations, you might make transparencies of student sheets that show grids like those on the computer screen, and use the overhead to show students the commands you would enter.

Investigations

Identifying Polygons

What Happens

Session 1: Is It a Polygon?   Students place shapes in two categories, polygons and not polygons, using a few examples as references. They list attributes of a polygon and features a polygon cannot have. Students find the polygons in a representation of the shapes in a Cubist painting, and for homework make their own polygon pictures.

Session 2: Making Polygons   Students are introduced to Power Polygons—shape pieces they will use throughout the unit. They use two or more shapes to make many polygons with different numbers of sides. For homework they draw polygons with vertices at intersections on a coordinate grid, for use in Session 3.

Session 3: Polygon Pictures with Coordinate Geometry   Students use coordinate geometry as a way of replicating polygon pictures drawn on a coordinate grid. They work together to list the coordinates of the points they used to draw their polygons. For homework, they work from a list of points to draw more polygon pictures on a coordinate grid.

Session 4: Coordinate Geometry with *Geo-Logo*   Students are introduced to the *Geo-Logo* commands jumpto and setxy as a way of making polygon pictures on the computer. Pairs trade pictures to draw on the screen, using the commands. Students pool what they know about the names of many-sided shapes, and, for homework, look for words with the same prefixes and for examples of polygon shapes in familiar objects.

Mathematical Emphasis

- Distinguishing between polygons and shapes that are not polygons
- Drawing polygons
- Locating points on a coordinate grid
- Using the *Geo-Logo* commands setxy and jumpto to draw polygons on the computer
- Recognizing and naming polygons by number of sides

What to Plan Ahead of Time

Materials

- Overhead projector and pen (all sessions)
- Chart paper (Session 1)
- Power Polygons: 1 bucket per 6–8 students (Session 2)
- Unlined paper: about 4 sheets per student (Sessions 2 and 3)
- Colored pencils, crayons, or markers (Session 2)
- Blank transparency (Session 2)
- Computers—Macintosh II or above, with 4 MB of internal memory (RAM) and Apple System Software 7.0 or later. Optimum: 1 per pair; minimum: 1 per 4–6 students. For fewer computers, see p. 4. (Session 4)
- Apple Macintosh disk, *Geo-Logo*™ for *Picturing Polygons* (Session 4)
- A large-screen monitor on one computer for whole-class viewing (recommended)

Other Preparation

- Duplicate student sheets and teaching resources (located at the end of this unit) as follows. If you have Student Activity Booklets, copy only the items marked with an asterisk.

For Session 1

Polygons and Other Figures* (p. 167): 1 transparency

Student Sheet 1, Is It a Polygon? (p. 162): 1 per pair

Student Sheet 2, Picasso's Polygons (p. 163): 1 transparency*, and 1 per student (homework)

Family letter* (p. 161): 1 per student. Remember to sign before copying.

For Session 2

Student Sheet 3, Coordinate Grids (p. 164): 1 transparency*, and 1 per student (homework)

For Session 3

Student Sheet 4, Hidden Polygon Pictures (p. 165): 1 per student (homework)

For Session 4

Student Sheet 5, Types of Polygons (p. 166): 1 per student (homework)

Geo-Logo User Sheet (p. 168): 1 to post at each computer*, and 1 per student

- Use the transparency of Student Sheet 3, Coordinate Grids, to make a sample House Picture (see p. 13). Note that each vertex of the polygon house must be on an intersection on the grid.
- Duplicate the Multiple Bingo materials (p. 196) as needed for use in Ten-Minute Math activities.
- If you plan to provide folders in which students will save their work for the entire unit, prepare these for distribution during Session 1.

Continued on next page

Computer Preparation

- Use the disk for *Picturing Polygons* to install *Geo-Logo* on each computer (See p. 156 in the *Geo-Logo* Teacher Tutorial.)

- Post the *Geo-Logo* User Sheet (p. 168) by each computer. This sheet provides information about running the program and entering commands.

- Work through the following sections of the *Geo-Logo* Teacher Tutorial:

- Plan how to manage the computer activities, depending on computer availability.

With five to eight computers: Follow the investigation structure as written. In Session 4, half the students work at the computers in pairs or threes while the other half work off computer. The groups then switch.

With a computer laboratory: Begin Session 4 with the computer activity. Students complete the off-computer activity as they finish their computer work.

With fewer than five computers: Introduce the on-computer activity in Session 4 and immediately assign some students to begin cycling through it. Make and post a schedule, assigning about 20 minutes of computer time for each pair of students throughout the day. Students may have to complete their computer work for this unit as you begin a new unit or while you engage the rest of the class in a short session.

Is It a Polygon?

What Happens

Students place shapes in two categories, polygons and not polygons, using a few examples as references. They list attributes of a polygon and features a polygon cannot have. Students find the polygons in a representation of the shapes in a Cubist painting, and for homework make their own polygon pictures. Student work focuses on:

- working with two-dimensional shapes
- distinguishing between polygons and shapes that are not polygons
- drawing polygons

Materials

- Overhead projector
- Polygons and Other Figures (transparency)
- Student Sheet 1 (1 per pair)
- Chart paper
- Student Sheet 2 (transparency, and 1 per student, homework)
- Family letter (1 per student)

Activity

Display the transparency of Polygons and Other Figures on the overhead. After students have had a few minutes to consider the transparency, distribute Student Sheet 1, Is It a Polygon? to each pair of students. Using the information on the transparency, pairs sort the shapes into two categories: those that are polygons and those that are not. On each shape they write *yes* or *no* (or some other notation) to keep track of their decisions.

While the students are working, draw a large circle on the board in which they can sketch polygons, and label it *Polygons*. Label an area outside of the circle *Not Polygons*. When pairs have finished Student Sheet 1, introduce the activity.

We are going to play a silent game. You will take turns coming to the board to draw one of the shapes from your sheet. Draw the shape *inside* the circle if you think it is a polygon, *outside* if it is not. Everyone else will think about the shape you drew. Is it in the right place?

If you think a shape is in the wrong place, you may put a question mark near it when you come up for your turn, instead of drawing a new shape.

Each pair selects one representative to go *silently* to the board and draw a shape from Student Sheet 1 either inside or outside the circle, or put a question mark by a shape they think has been placed incorrectly. To emphasize the game's silence, you might gesture to indicate turns instead of calling names.

Picking Out Polygons

Once all the shapes have been placed, allow some time to discuss the placements. Disagreements offer particularly good opportunities for students to explain and defend their reasoning about what qualifies as a polygon.

Unless there are several disagreements, there will probably not be enough shapes for each pair to draw one on the board. If there are students who do not get a chance in this activity, ask them to offer their thoughts at the beginning of the next one.

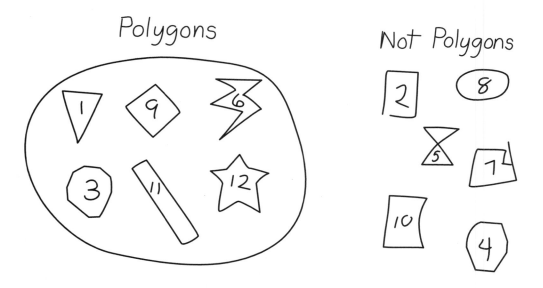

Defining a Polygon

In the silent drawing game, students *showed* their developing sense of what a polygon is and is not. Now help them verbalize that sense by asking them to describe the attributes of a polygon.

For a shape to be a polygon, what must be true? What cannot be true?

List students' ideas on chart paper. You might want to begin two sentences and ask students to finish them, for example, "Polygons *must* have…" and "Polygons *cannot* have…." The **Dialogue Box,** Describing Polygons (p. 8), illustrates the list one class compiled.

As an extension, suggest that students write their own definition of a polygon.

Polygon Art

Save the last few minutes of the session to introduce the homework: drawing a picture made up entirely of polygons. Display the transparency of Student Sheet 2, Picasso's Polygons, on the overhead. Briefly explain that in the style of art we call *cubism,* artists like Pablo Picasso represented people and objects by breaking them down into many flat shapes.

This picture shows the shapes in a drawing done by Picasso in 1918, *Woman in an Armchair.* (The original drawing includes some curved lines; all are represented by straight lines here to emphasize the polygonal forms.) If possible, show prints of other cubist works by Picasso.

Discuss what various polygons in the drawing might represent. Some students may comment that the picture as a whole is not a polygon because of the many interior lines. Acknowledge that they are correctly applying the definition of a polygon, and redirect their attention to the smaller polygons that make up the whole picture.

Do you notice any shapes you think are *not* polygons?

Many students say that the larger shapes surrounding the eyes in the picture are not polygons because they have other shapes in the middle (the eyes). Decide as a class what rules to follow in their own drawings.

Session 1 Follow-Up

Homework

Picasso's Polygons Send home the family letter or the *Investigations* at Home booklet. In addition, students take home their own copy of Student Sheet 2, Picasso's Polygons. Students may draw their own picture on the back. Suggest that they try drawing some things that have natural curves, such as people, animals, or plants. Encourage students to use polygons with many different numbers of sides, not just triangles and quadrilaterals. Suggest that they work in pencil so they can erase, and remind them to follow the rules for polygons.

Describing Polygons

These students are describing attributes that make a figure a polygon. The teacher is recording some of their ideas on a list.

Antonio: It can't be a circle or have curved lines.

[*Teacher writes:* It can't have curved lines.]

Trevor: Yeah, it has to have corners or points.

[*Teacher writes:* It has corners (vertices).]

Shakita: Do they have to have four or more sides?

Based on the polygons we've seen, is that a reasonable attribute?

Heather: Is a triangle a polygon?

Cara: Yes, we should make it less. Two or more sides.

Duc: I never heard of a shape that has two sides.

Cara: OK, three or more sides.

[*Teacher writes:* It has three or more sides.]

What other properties does a polygon have to have?

Toshi: They have to have a line all the way around.

[*Tracing the outline of one of the polygons*] **This is called a *closed* shape. What do you think that means?**

Alani: There's no open place at the edge. It's like a dog inside a fenced yard.

[*Teacher writes:* It is a closed shape.]

Matt: Another thing is it doesn't cross over. The lines can't meet in the middle.

Maricel: I know what you mean. It can't be an 8 shape. It only has one space in it.

We call it a simple shape. It is just one space.

Greg: Cut them in half and they're even. Like it has to be the same after you fold it. If you fold it in half and if it's the same shape still, then it's a polygon.

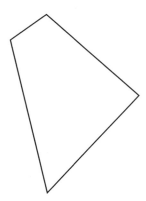

What about this shape?

What do others think? Does a polygon need to fold in half exactly?

Natalie: No, it can be weird and still be a polygon.

The class ends up with the following list:

Polygons

Must be true:	Cannot be true:
■ It has only straight sides.	■ It doesn't cross over.
■ It has three or more sides.	■ It can't have curved lines or be a circle.
■ It has corners (vertices).	■ It can't have more than one space in it.
■ Lengths of sides don't have to be equal.	
■ It is a closed shape.	
■ It doesn't have to be a typical shape (like a square or rectangle) or a symmetrical shape that can be folded in half exactly.	

Making Polygons

What Happens

Students are introduced to Power Polygons—shape pieces they will use throughout the unit. They use two or more shapes to make many polygons with different numbers of sides. For homework they draw polygons with vertices at intersections on a coordinate grid, for use in Session 3. Their work focuses on:

- working with two-dimensional shapes
- recognizing polygons by number of sides
- creating and drawing polygons

Materials

- Power Polygons (1 bucket per 6–8 students)
- Overhead projector
- Blank transparency and overhead pen
- Unlined paper (about 2 sheets per student)
- Colored pencils, markers, or crayons
- Prepared transparency of House Picture
- Student Sheet 3 (1 per student, homework)

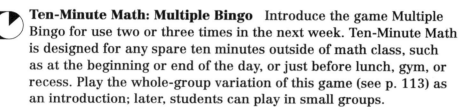 **Ten-Minute Math: Multiple Bingo**   Introduce the game Multiple Bingo for use two or three times in the next week. Ten-Minute Math is designed for any spare ten minutes outside of math class, such as at the beginning or end of the day, or just before lunch, gym, or recess. Play the whole-group variation of this game (see p. 113) as an introduction; later, students can play in small groups.

Each player will need a 100 chart and a crayon or marker. Calculators are optional. Players take turns drawing a card from the Multiple Bingo deck and calling out the number. Every player finds a *multiple* of that number on his or her 100 chart and marks or colors that square. Players write the number that was called in their marked square so they can check their accuracy later.

The player who draws a Wild Card decides on the number to be used. The best strategy is to choose a number that helps the player's own game but doesn't help others.

The game continues until a player has marked five numbers in a row for Bingo.

For a full description and other variations on this game, see p. 112.

Putting Together Power Polygons

Hold up various Power Polygon shape pieces and begin assembling two or more pieces in a shape on a blank transparency on the overhead. It's fine to make a nonregular polygon for this demonstration, but try to make one with fewer than seven sides so they will be easy to count.

As we investigate polygons, we are going to put together shape pieces like these to make polygons of all shapes and sizes.

Remembering the definition of a polygon they just developed, some students will protest that the figure you are making is not a polygon because it has "more than one space in it," because it has interior lines where the shape pieces are put together. Acknowledge that they are right, but explain that for the purposes of the activities in this unit, you are all going to ignore the interior lines. You may want to use the pieces to demonstrate a shape that is clearly *not* a polygon, to clarify that most of the rules still pertain.

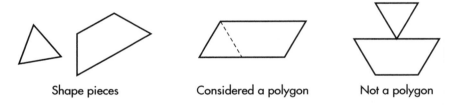

Shape pieces   Considered a polygon   Not a polygon

Suggest that whenever students make polygons from more than one shape piece, they trace the outer edges darkly to show the shape, while tracing very lightly or using only dotted lines to show the interior pieces that make up the shape.

When you have completed a figure, ask students to count the sides.

I have put these pieces together to make a polygon. Does everyone agree that it is a polygon? How many sides does it have? Check with your partner—do you agree?

If there is any disagreement about the number of sides, invite two students who differ to come to the overhead and show how they counted. Be on the lookout for students who count the number of pieces that form a side, instead of the actual number of sides; a single straight side may be formed by more than one piece.

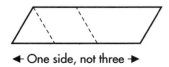

◀ One side, not three ▶

Hold the pieces flat, one or two at a time, as you trace around them with an overhead pen. Then mark the lines in the interior of the polygon with dotted lines, either sliding the entire figure off the drawing and marking the divisions, or removing one piece at a time and tracing the exposed edge with a dotted line. Labeling each interior piece with the letter of the shape will help students begin to recognize the variety of shapes in the set. Also label the exterior of the polygon by numbering each side as you count it aloud, and write the total near it.

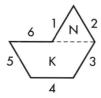

You will be working in small groups to make polygons with shape pieces like these. Try to put together pieces with different letters to make one shape. After you make a polygon, trace it on paper, just like I did on the overhead, and label it.

Each group has three tasks:

1. **Make many different three-sided shapes.**
2. **Make many different four-sided shapes.**
3. **Make shapes with more than four and up to ten sides. Your group needs to find only one or two of the shapes with five to ten sides.**

Write the three parts of the assignment on the board. Emphasize that their main challenge lies in tasks 1 and 2—finding as many really different three- and four-sided polygons as they can. Distribute the sets of Power Polygons, each bucket to be shared by six to eight students. If this is your students' first experience with Power Polygons, allow time for them to investigate and explore the shape pieces.

As you observe the students at work, comment on the shapes they are creating. Use proper terminology, as described in the **Teacher Note,** Beyond Vocabulary (p. 14). For example:

I see you have made a … [*count the sides and point to them*] 1, 2, 3, 4 … a *quadrilateral.* Look at how this part of it goes in. That means it's *concave,* as if this part of the shape caved in.

This shape is called a *chevron.*

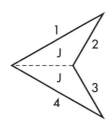

Here's a shape that is a special kind of quadrilateral called a _trapezoid_. It has one pair of parallel sides *[point to them].*

In this way you introduce new vocabulary naturally. Refer to the **Teacher Notes,** Types of Polygons (p. 27) and Classification of Triangles and Quadrilaterals (p. 42), for the names and categories of different shapes.

Some students may be startled when you call a nonregular polygon by the same name used for a regular polygon. For example, they may think that a hexagon has to have six equal sides and six equal angles. Explain that the name of any polygon tells only how many sides it has. *Any* polygon with four sides is a quadrilateral. *All* pentagons have five sides. *Any* three-sided polygon is a triangle. To identify a polygon that has equal sides and equal angles, we must call it a *regular* polygon.

Hexagons

Regular hexagon

As a challenge, groups who finish early can try to make *convex* shapes with four to ten sides. A shape is *convex* if all its angles are smaller than 180°. A trapezoid is a convex quadrilateral; a chevron is a *concave* quadrilateral. By mathematical definition, a shape is convex if we can draw a straight line between any two points in the shape without going outside of the shape. In a concave shape, the lines between some points go outside the polygon.

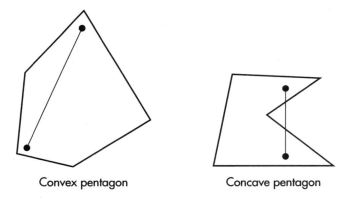

Convex pentagon   Concave pentagon

Toward the end of the session, students choose a favorite shape or a page of shapes to trace, color in, and post. Encourage them to label the shapes if they know the names.

Reserve the last few minutes of class to introduce the homework in which students draw a polygon picture on a coordinate grid. Display your House Picture transparency as an example.

Polygon Pictures

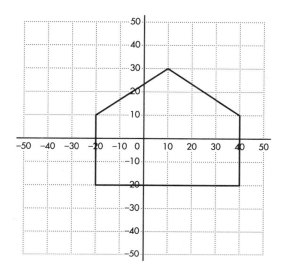

Explain that the corners, or vertices, of their polygons must be at intersections on the grid, as in the House Picture, because later they are going to tell the computer how to draw the shapes. Also suggest that they keep their drawings simple, as too many lines or points will take too long to enter on the computer.

Save your House Picture transparency for further use in the next session.

Session 2 Follow-Up

Coordinate Grids Students take home a copy of Student Sheet 3, Coordinate Grids, and draw a simple polygon picture that has all its vertices at intersections of the grid. They draw their picture on the top grid, saving the bottom grid for use in the next session.

Homework

The work in this unit will bring up many vocabulary words that students do not yet know. Avoid the temptation to stop to teach a lesson specifically on vocabulary or to test students on vocabulary words. This temptation is especially great in the beginning of Investigation 2, when students are introduced to a large amount of vocabulary as they try to describe polygons by the number of sides, and to describe kinds and characteristics of triangles and quadrilaterals.

Throughout the unit, you and your students will be using words that help you describe different kinds of polygons and angles. While students are working together with polygons on and off the computer, they will talk about the shapes in their own words. For example, a "pointy angle" may mean an acute angle and a "wide angle" an obtuse one. However, the activities in this unit will motivate students to use more precise language. As students work to make themselves clear, you will hear some unusual, but accurate, descriptions. Try to achieve a balance between using the most accurate words and agreeing on classroom vocabulary that makes the most sense to students.

Students will learn the mathematical terms as they hear them in context. Use the correct terms yourself and informally explain to students what they mean. As students hear you and others use the terms, they will adopt the ones that make sense to them—but this process will take time. Do not expect students to use terms as soon as they hear them. The activities in this unit provide repeated experience with the same terms and concepts, so students will have a number of opportunities to pick up the new vocabulary. Be aware that some students will adopt it easily, while others will need more time and exposure.

More important is that students develop accurate concepts for shapes for which they already have names. Many 10-year-olds (and even older students) have very limited ideas about common shapes. For example, they may call a rectangle a square; they believe that a triangle must have symmetry (must have at least two equal sides) and must not be tilted on the page. Help students build concepts without giving them definitions. Allow them to try out their ideas in class discussions, but expect them to make themselves clear and to justify their conjectures.

Polygon Pictures with Coordinate Geometry

What Happens

Students use coordinate geometry as a way of replicating polygon pictures drawn on a coordinate grid. They work together to list the coordinates of the points they used to draw their polygons. For homework, they work from a list of points to draw more polygon pictures on a coordinate grid. Students' work focuses on:

- locating points on a coordinate grid
- creating and drawing polygons

Materials

- Overhead projector and pen
- House Picture transparency
- Unlined paper (about 2 sheets per student)
- Students' polygon pictures on Student Sheet 3 (from their homework)
- Student Sheet 4 (1 per student, homework)

Activity

Finding the Coordinates

Note: Students who have done the *Investigations* grade 4 unit, *Sunken Ships and Grid Patterns,* will be familiar with coordinate grids and how to identify points on them. Enlist their help as you review the procedure for others in the class.

On the overhead, show the House Picture transparency.

A coordinate grid is made with two crossed number lines that we call the x-axis and the y-axis. *[Point to the axes as you name them.]* **The two lines or axes cross at the zero points.**

If we look at the horizontal axis, or x-axis, we see the numbers that show how far to the side a point is located. The numbers going to the right of zero are positive numbers, and the numbers going to the left are negative.

If we look at the vertical axis, or y-axis, we see the numbers that show how far up or down a point is. The numbers going up from zero are positive numbers, and the numbers going down from zero are negative, just like on a thermometer.

We can use these numbers to give directions for drawing a picture, like this house, on a grid. We'll go around the picture and name each of the vertices (points, corners), in order, with two numbers. The first number tells how far to the left or right of zero the point is, and the second number tells how far up or down the point is. These numbers together are called *an ordered pair*. We always write the *x*-coordinate before the *y*-coordinate.

Point to a vertex of the house and show how to name that point: Trace with a finger up or down to the *x*-axis to find its *x*-coordinate, and trace across to the *y*-axis to find the *y*-coordinate. Name the coordinates of the point and write the ordered pair—for example, (–20, 10)—on the transparency next to the point, and start a listing of coordinates in the space to the right of the grid. Moving clockwise, point to the next vertex.

What is the pair of numbers for this point?

When students agree, write the ordered pair for this point under the first ordered pair.

On scratch paper, students copy down the first two points and then figure out the name (the coordinates) of the third point and write it down. They share what they think it is. Circulate to see how students are naming the point. When you feel they are ready, have them find the coordinates for the rest of the vertices by themselves, consulting with neighbors to compare their list of points and discuss any discrepancies. Encourage students to teach one another as they work, so that you offer advice only when a whole group requests it.

As the students report the remaining coordinates for the polygon, record them on the transparency. Point out that by connecting these points in order, they could draw the same picture on *any* coordinate grid. Save the

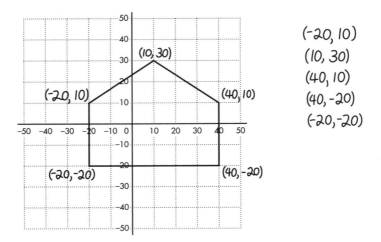

$$(-20, 10)$$
$$(10, 30)$$
$$(40, 10)$$
$$(40, -20)$$
$$(-20, -20)$$

Teacher Checkpoint

Identifying Coordinates

House Picture transparency for further discussion in the next session.

Students take out their homework with the polygons they drew on Student Sheet 3. Working individually or in pairs, they list the coordinates of the points in their polygons next to the grid. Circulate quickly to see that students can label points accurately. Use this activity as a checkpoint for students' ability to use coordinates. The **Teacher Note,** Putting It All Together (p. 18), lists some common misconceptions to watch for.

Students who finish early can create a new polygon picture on the second grid of the student sheet, writing the coordinates on a separate piece of paper. They trade pictures with a partner, list the coordinates of each other's polygons, then compare to see if they agree. They save these sheets for their first work on the computer in Session 4.

Students with little background in coordinate geometry may need further practice with reading and locating points on a grid. You may want to take time now to have students create and write the coordinates for more polygon pictures to practice finding and naming points.

Session 3 Follow-Up

Hidden Polygon Pictures   Students take home a copy of Student Sheet 4, Hidden Polygon Pictures, and use the points listed to draw the polygons on the coordinate grid.

Students often have difficulty taking the two numbers that make up one coordinate, such as (10, 30), and thinking of them as one point or location. You may see students doing the following during the off-computer activities:

■ Referring to a point on the *x*-axis as 50 instead of (50, 0).

■ When asked for coordinates for two points, giving all the numbers in a row, such as 10 0 30 10, without being able to say which numbers are paired.

■ Referring to the origin (the intersection of the axes) as 0 (zero) instead of (0, 0).

■ Being confused about how the two numbers relate; for example, referring to the point (8, 9) as eight-ninths.

During the on-computer activities, you may see students doing the following:

■ Believing that `setxy [10 30]` would move the turtle over 10 and up 30 from its current position instead of moving it to the point (10, 30) relative to the origin.

■ Claiming that the length of the oblique line drawn by the two commands `setxy [0 0]` `setxy [50 50]` is 50 (or 100).

In many paper-and-pencil tasks and tests, students get by, even though they have misconceptions like these. When students use the computer language *Geo-Logo,* however, they receive feedback that alerts them to problems. Thus, one strategy for dealing with these misconceptions is to have students design their pictures off computer and then write the commands for them on the computer. When they try to have *Geo-Logo* draw the picture, they receive feedback on their ideas. They should then discuss which commands didn't work and figure out how to change their approach.

Similarly, students can exchange coordinates for a picture to check each other's ideas.

General discussion might emphasize that two numbers are necessary to locate any one point.

Coordinate Geometry with *Geo-Logo*

What Happens

Students are introduced to the *Geo-Logo* commands `jumpto` and `setxy` as a way of making polygon pictures on the computer. Pairs trade pictures to draw on the screen, using the commands. Students pool what they know about the names of many-sided shapes, and, for homework, look for words with the same prefixes and for examples of polygon shapes in familiar objects. Their work focuses on:

- using the *Geo-Logo* commands `jumpto` and `setxy` to write a procedure for drawing a polygon on the computer
- locating points on a coordinate grid
- creating and drawing polygons

Materials

- Completed homework (Student Sheet 4)
- Overhead projector and pen
- House Picture transparency
- Computers with *Geo-Logo* for *Picturing Polygons* installed
- Projection device for computer (optional)
- *Geo-Logo* User Sheet (1 per computer and 1 per student)
- Students' polygon pictures and points on Student Sheet 3
- Student Sheet 5 (1 per student, homework)

Activity

Take a few minutes to review the homework on Student Sheet 4, Hidden Polygon Pictures, and make sure everybody "found" the pictures of the arrow and the tugboat. Assure students who had difficulty that they can work a few minutes with other students in off-computer time later to find their mistakes and review how to interpret the ordered pairs of coordinates.

Hidden Picture Homework

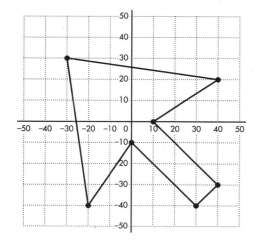

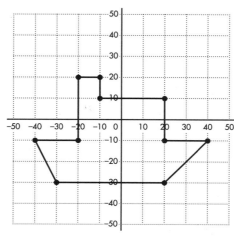

Writing *Geo-Logo* Commands

Note: It is critical that you have worked through the first sections of the *Geo-Logo* Teacher Tutorial (Overview, Getting Started with *Geo-Logo,* and Polygons with Coordinates, pp. 117–119) before getting students started on the computer activities in this unit.

Students who have done the *Investigations* grade 3 unit, *Turtle Paths,* or the grade 4 unit, *Sunken Ships and Grid Patterns,* have some experience with *Geo-Logo* commands and procedures. If your students have not been exposed to *Logo* before, you may need to spend more time demonstrating the basic *Geo-Logo* commands and procedures. The *Geo-Logo* Teacher Tutorial (p. 117) includes an overview of the basic commands. If your students are familiar with *Geo-Logo,* keep this introduction to setxy and jumpto brief so they will all have time to work on the computer.

Explain that in this unit, the students will be using computers and the computer program *Geo-Logo* for several activities. You might want to go over your classroom work procedures for the unit; the **Teacher Note,** Introducing Work Procedures on the Computer (p. 25), offers some ideas.

Display the House Picture transparency, now labeled with ordered pairs, on the overhead projector as you introduce the procedure for writing commands to draw the same picture using *Geo-Logo.*

We talked earlier about how we could give directions for drawing this picture by listing the points and telling someone to connect them in order. We can also draw pictures this way on the computer, using *Geo-Logo.* We have to write commands that the computer understands, and that means changing slightly how we write ordered pairs.

In *Geo-Logo,* whenever you open an activity with a grid, the turtle will be sitting on the point (0, 0). This is fine, if that's where you want to start your drawing. If you want to start at a different point, you use the jumpto command.

Write the heading *Geo-Logo Commands* next to the bottom grid on the transparency, and write the command jumpto under it.

The jumpto command makes the turtle jump to a specific point on the grid. We can make it start at our first point by typing jumpto *[name the coordinates],* or we can start at a different vertex.

Ask the class to agree on a first point, using a vertex of your House Picture. Write the complete *Geo-Logo* command on the transparency; for example, jumpto [-20 10]. Put a dot on the second grid at that point to show where the turtle will start.

Make sure students understand that to change an ordered pair from standard notation to a *Geo-Logo* command, we use brackets instead of parentheses, and no comma between the two numbers. It is important to leave a space, however, so the computer knows where the first number ends and the second begins. This is discussed further in the **Teacher Note,** Notational Conventions in *Geo-Logo* (p. 26).

What is the next point we want to go to in the drawing? *[Take students' suggestions.]* **The command that tells the computer to draw a line between two points is** setxy *[pronounce as "set x, y"].* **If we enter** setxy [10 30], **for example, the turtle will move to the point (10, 30), drawing a line as it goes.**

Write the setxy command on the transparency under the jumpto command, and a draw a line on the grid to show how the turtle would move.

List one or two more *Geo-Logo* commands that will continue the picture, taking students' suggestions for exactly how to write them and drawing the lines the turtle would draw. When students are ready, they copy the first commands on a sheet of scratch paper and write the rest of the commands to finish the drawing using setxy. When students have finished, ask for their commands and record them on the transparency.

Classroom Management

After the off-computer introduction to jumpto and setxy, introduce the off-computer activity, Brainstorming the Names of Polygons, to the entire class. Explain that everyone will work on this activity at some point during this session. If you have a projection device or a large enough monitor to demonstrate the basics of your computer and of *Geo-Logo* to the class, explain the on-computer activity to the whole class next. Otherwise, once the off-computer task has been explained, half the students can begin working on it while you introduce the on-computer activity, Making Polygon Pictures with *Geo-Logo,* to the rest of the class.

Students spend half of the remaining time in Session 4 doing either the on-computer activity or the off-computer activity, and then switch. At that time, if you have not already done so, you will need to introduce the computer activity and procedures to the new on-computer group.

Off Computer

Brainstorming the Names of Polygons

Distribute Student Sheet 5, Types of Polygons. Students fill in the names of polygons with different numbers of sides and begin to write related words in the third column. Remind students the prefixes should give the word a number meaning. For example, the prefix *dec-* means "ten" in the word *decade,* which means "ten years." Students may also see the letters *dec* at the beginning of the words *decal* and *declare,* but the meanings of those words are not related to the number ten.

Students can work together to pool any information they already know or have found. Each student records what his or her group finds on a list, then takes the list home and continues working on it for homework. See the **Teacher Note,** Types of Polygons (p. 27), for a list of ideas accumulated from several classrooms.

❖ **Tip for the Linguistically Diverse Classroom**   Pair English-proficient students with others to complete this task. Partners draw a picture next to each word they list.

On Computer

Making Polygon Pictures with *Geo-Logo*

Gather students around the largest computer display available. You can make the typeface larger for better viewing by choosing **All Large** from the **Font** menu. When you have finished demonstrating, you can return the font to its regular size by choosing **All Small** from the **Font** menu.

Distribute copies of the *Geo-Logo* User Sheet and explain its use. Post one copy by each computer and encourage students to keep a copy in their folders to use as a reference.

Demonstrate as necessary:

■ How to turn on the computer.

■ How to open *Geo-Logo* by double-clicking on the icon.

■ How to open the Polygons with Coordinates activity by clicking on it once.

To prepare students for this activity, briefly do the following:

■ Demonstrate how to enter commands by typing them in the Command Center. As an example, draw the house picture, beginning with your initial `jumpto` command and the first `setxy` command. (An abbreviation for `jumpto`, `jt`, can be a useful shortcut.)

- Show how to edit commands using the mouse, the **<delete>** key, and the arrow keys to correct mistakes made, or to change the numbers in the brackets.

- As you enter the procedure to draw the house, show students how to use the **Copy** and **Paste** commands. For example:

  Highlight `setxy`.

  Choose **Copy** from the **Edit** menu (or enter ⌘C).

  Click where you would like that material to reappear.

  Choose **Paste** from the **Edit** menu (or enter ⌘V).

  This way, you avoid having to retype `setxy` every time you enter a new command. You can use **Paste** again and again to get several copies. Hit **<return>** after each **Copy** command to go to a new line.

- Specifically demonstrate the use of three tools on the tool bar:

 Erase One   Erase All   Teach

  Use Erase One to erase the last command entered.

  Use Erase All to erase *everything* in the Command Center. Students should use the Erase All tool whenever they want to start a new polygon picture. (They can also highlight material to be erased and use the **<delete>** key, then hit the **<return>** key to see the changes in their picture.)

  Use Teach to define the commands as a procedure. Enter a name in the dialogue box—for example, a name describing your picture, in this case house—then enter that name in the Command Center and hit **<return>** to run the procedure.

When you have finished entering the commands, compare the computer drawing to the one on the transparency.

After your demonstration, students work together at the computer to write procedures to draw polygons with *Geo-Logo*. They start with the polygon pictures they made on Student Sheet 3. If there is time, students may try to write a procedure to draw another polygon picture.

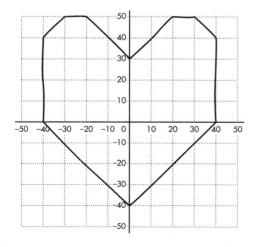

Jump to [0, -40]
set xy [40, -0]
set xy [40, 40]
set xy [30, 50]
set xy [20, 50]
set xy [0, 30]
set xy [-20, 50]
set xy [-30, 50]
set xy [-40, 40]
set xy [-40, 0]
set xy [0, -40]

Pooling Polygon Names

A few minutes before the end of the session, come together as a group to pool all the names for polygons with different numbers of sides that your students found during their off-computer work. Provide any that they have not yet found, up to 12 sides, including *hendecagon* (11 sides) and *dodecagon* (12 sides).

Students may then guess that a 13-sided polygon is a tridecagon and a 14-sided polygon is a quaddecagon. Compliment their logical reasoning, but explain that after 12 sides, there is another convention for naming polygons—by the number of sides they have.

A polygon with 14 sides is called a 14-gon. What do you think a polygon with 13 sides is called? How about one with 23 sides?

Start a class chart similar to the one shown in the **Teacher Note,** Types of Polygons (p. 27). List the names of the polygons with up to 12 sides and leave an open column for related words and visual examples that the students will be locating as homework.

Students write the names of polygons with 3 to 12 sides to take home as reference for this homework.

Session 4 Follow-Up

Homework

Types of Polygons Students continue to fill in the third column of Student Sheet 5, Types of Polygons, with other words that have the same prefixes as the polygon names and write the meanings of the words. They also look for examples of the polygon shapes in familiar objects (for example, a pencil in cross section is a hexagon, and a stop sign is an octagon) and list them on a separate piece of paper.

❖ **Tip for the Linguistically Diverse Classroom** Students draw pictures of familiar objects with different polygonal shapes.

As students come into class the next day, they add their findings to the class chart, checking first to see that each example they are adding isn't already listed.

Introducing Work Procedures on the Computer

During this unit, students will be doing activities both on the computer and at their desks. Assuming there are not enough computers for everyone to use at once, students will have to take turns. You will need to explain how the class will operate so that everyone has an equal opportunity to do the computer work. (Managing the Computer Activities in This Unit, p. I-21, offers some practical suggestions.)

Following are some procedures you may want to establish and explain to students:

■ For computer activities, the math session (or two consecutive sessions) will be divided into two equal blocks of work time. Half the students will work on the computer during the first block of time while the rest of the class does off-computer activities; then the groups will switch.

■ Both the on- and off-computer activities will be explained to everyone at the beginning of each class session.

■ Students will work with the same partner(s) for the entire unit. Sometimes two pairs will work together.

■ Pairs of students will use the same computer throughout the unit or save their work on a disk. Note that if students are working on computers with screens of different sizes, there may be times when a procedure written on a computer with a large screen will not fit on a smaller screen.

■ When working at computers, partners take turns, with one person using the mouse and the keyboard while the other reads and writes. They trade places after a given period of time, for example, every 5 minutes.

■ If students have questions about how to proceed, they first check the *Geo-Logo* User Sheet posted at the computer. If that doesn't help, they ask someone who has more experience with the computer or someone who has already done the computer activity.

Students should understand that the conventions for specifying coordinates in *Geo-Logo* differ slightly from traditional mathematical notation. For example:

In mathematics: (–20, 60)
In *Geo-Logo*:   [-20 60]

For the `jumpto` and `setxy` commands, we use square brackets in *Geo-Logo* instead of parentheses, and there is a space but no comma between the coordinates.

Discuss with your students that many such conventions in mathematics and computer programming are arbitrary—they are "right" simply because people have agreed to do things that way, for the sake of consistency. List on the board some other examples of mathematical conventions—things that people have agreed upon. For example:

the symbol + for *add,* and – for *subtract*

calling the horizontal axis *x* and the vertical axis *y*

writing the *x*-coordinate before the *y*-coordinate in an ordered pair

Why aren't the computer conventions the same as the traditional mathematical conventions? It has to do with how the computer interprets certain symbols. For example, *Geo-Logo* uses brackets for *any* list of items, not just for coordinates. Other bracketed items in *Geo-Logo* include repeat statements and print statements :

repeat 4 [fd 50 rt 90]
print [This is my drawing.]

To make the design clean, simple, and elegant, all lists in *Geo-Logo* use this convention.

The computer uses parentheses, on the other hand, to mean "operate on this as a unit." For example:

print 5 * (3 + 4)

means "add the 3 and 4 before you multiply by 5." In this case, mathematicians use the same convention as the computer.

It's easy for humans to decide that one set of parentheses means "operate on this as a unit" and another designates an ordered pair. We use the context to tell us what is intended. But the computer cannot do this, so some new conventions are needed to make the computer instructions completely unambiguous.

Types of Polygons

In Session 4, students start a list of the names of polygons with up to 12 sides and look for words that contain the same numerical prefixes. They also look for examples of polygon shapes in familiar objects. These ideas were collected by several fifth grade classes.

Objects with polygonal shapes:

home plate (5-sided)
pencil (6-sided)
beehive cells (6-sided)
one-way sign arrow (7-sided)
stop sign (8-sided)
starfish (10-sided)
Susan B. Anthony dollar (11-sided)

Number of sides	*Name of polygon*	*Related words or objects*
3	triangle	triceratops (dinosaur with 3 horns) trillium (plant with 3 leaves) trimester ($\frac{1}{3}$ of a school year) triceps (muscle connected in 3 places) triathlon (athletic contest with 3 events) tricycle (3-wheeler) trident (spear with 3 prongs) trilingual (speaking 3 languages) triennial (every 3 years) trilogy (3 books that go together) triple (3-base hit or multiply by 3) triplets (3 babies) trio (3 musicians) triangle (musical instrument with 3 sides) tripod (stand with 3 legs) triad (group of 3)
4	quadrilateral	quadragenarian (4 decades old) quadruplets (4 babies) quartet (group of 4) quad or quadrangle (area enclosed on 4 sides) quadruple (multiply by 4) quadruped (animal with 4 legs) quadriceps (4-part muscle) quadrennial (every 4 years) quadrant ($\frac{1}{4}$ of a circle) quarter ($\frac{1}{4}$ of $1) quart ($\frac{1}{4}$ of a gallon)
5	pentagon	pentathlon (athletic contest with 5 events) pentangular (having 5 angles) pentastitch (poem with 5 lines) Pentagon (5-sided building) pentameter (line of poetry with 5 beats)
6	hexagon	hexachord (6-tone musical chord) hexameter (line of poetry with 6 beats) hexapod (6-legged animal; insect)
7	heptagon or septagon	septet (group of 7) September (7th month of Roman calendar) septuagenarian (7 decades old)
8	octagon	octave (musical interval of 8 tones) octopus (sea animal with 8 legs) octet (group of 8) October (8th month of Roman calendar)
9	nonagon	nonagenarian (9 decades old) November (9th month of Roman calendar)
10	decagon	decathlon (athletic contest with 10 events) decimal (base 10 system) decade (10 years) December (10th month of Roman calendar) decimeter ($\frac{1}{10}$ of a meter) decaliter (10 liters)
11	hendecagon	hendecasyllabic (having 11 syllables)
12	dodecagon	dodecahedron (solid figure with 12 faces) dodecasyllabic (having 12 syllables) Dodecanese (group of 12 Greek islands)

Triangles and Quadrilaterals

What Happens

Sessions 1, 2, and 3: Sorting Polygons   After a brief homework review, students begin sorting polygons: the three-sided figures and four-sided figures from the Guess My Rule Cards. They think of different ways to categorize triangles and quadrilaterals and then complete the statements "All triangles...," "Some triangles...," "All quadrilaterals...," and "Some quadrilaterals...." Finally they play Guess My Rule with all the cards.

Sessions 4 and 5: Making Shapes That Follow Rules   Using Power Polygons off the computer and *Geo-Logo*'s coordinate commands jumpto and setxy on the computer, students make triangles and quadrilaterals that fit given descriptions. They then discuss which of the shapes are impossible to make and which are difficult to make.

Sessions 6 and 7: Using Move and Turn Commands   Students work with *Geo-Logo* move commands (fd and bk) and turn commands (rt and lt). They write procedures that draw a square and a rectangle, and see how they can use the repeat command to write the same procedures in a shorter form. On computer, students use *Geo-Logo*'s move and turn commands to draw an equilateral triangle (and, if time permits, other shapes). In an off-computer assessment, students answer questions about the hierarchical categories of polygons; for example, why a square is also a rhombus. As a follow-up to the computer work, students discuss the difference between turns and angles.

Session 8: Finding Angle Sizes   Students use a number of different strategies to find the sizes of angles in the Power Polygons. They discuss their strategies, and they learn to use Turtle Turners as a way of checking their accuracy. For homework, students draw angles by estimating the size, then check each one with a Turtle Turner.

Session 9: Angles and Turns Together
Students play a game on the computer that relates the size of an angle to the size of the turn made to form that angle (the supplement of the angle). Off the computer, they estimate and draw some of the angles in the Power Polygons sets, and they use Turtle Turners to measure turns and angles in triangles.

Mathematical Emphasis

- Reasoning and communicating about properties of geometric shapes
- Sorting and classifying triangles and quadrilaterals
- Developing vocabulary to describe special triangles and quadrilaterals
- Generating geometric figures from descriptions of their properties
- Estimating and measuring the size of angles and turns

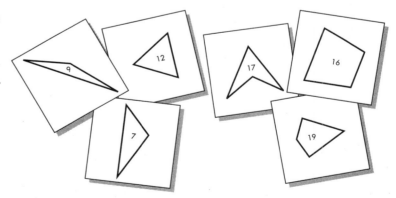

What to Plan Ahead of Time

Materials

- Power Polygons: 1 bucket per 6–8 students (Sessions 1–5 and 8–9)
- Envelopes or resealable plastic bags for storing decks of cards: 1 per student (Sessions 1–3)
- Rubber bands or paper clips for sorted cards: 1 per pair (Sessions 1–3)
- Loop of string that encloses about half the space on the overhead, or a blank transparency (Sessions 1–3)
- Chart paper (Sessions 1–3 and 6–7)
- Overhead projector and pen (Sessions 1–8)
- Blank transparencies (Sessions 6–8)
- Computers (Sessions 4–7 and 9)
- Rulers: 1 per 2–3 students (Session 9)

Other Preparation

- Duplicate student sheets and teaching resources (located at the end of this unit) as follows. If you have Student Activity Booklets, copy only the items marked with an asterisk.

For Sessions 1–3

Guess My Rule Cards (p. 182): 1 deck per pair (preferably on card stock); 1 transparent set*; and 1 set per student (homework)

Student Sheet 6, Is Every Three-Sided Polygon a Triangle? (p. 169): 1 per student (homework)

Student Sheet 7, Is Every Square a Rectangle? Is Every Rectangle a Square? (p. 170): 1 per student (homework)

Student Sheet 8, How to Play Guess My Rule with Shapes (p. 171): 1 per student (homework)

For Sessions 4–5

Student Sheet 3, Coordinate Grids (p. 164): available in class; 2 per student (homework)

Student Sheet 9, Can You Make These Triangles? (p. 172): 1 per student, and 1 transparency*

Student Sheet 10, Can You Make These Quadrilaterals? (p. 173): 1 per student, and 1 transparency*

Student Sheet 11, Find the Fourth Vertex (p. 174): 1 per student (homework)

For Sessions 6–7

Student Sheet 12, Some Shapes Fit Many Categories (p. 175): 1 per student

Student Sheet 13, What Shape Does It Draw? (p. 176): 1 per student (homework)

For Session 8

Student Sheet 14, Angles in the Power Polygons (p. 177): 1 per student

Turtle Turners* (p. 184): 1 transparency per 2 students

Student Sheet 15, Estimating Angles (p. 179): 1 per student (homework)

For Session 9

Student Sheet 16, Angles and Turns (p. 180): 1 per student

Student Sheet 17, What Do You Know About 45° and 60° Angles? (p. 181): 1 per student (homework)

Continued on next page

■ Before starting the investigation, read the **Teacher Note,** Classification of Triangles and Quadrilaterals (p. 42).

■ Before Session 1, enlist students' help in cutting apart the Guess My Rule Cards. Place each deck in an envelope or plastic bag. Also cut apart the transparent cards.

■ Before Session 8, cut apart the Turtle Turners transparency to give each student one Turtle Turner for class and one for homework.

Computer Preparation

■ Be sure a copy of the *Geo-Logo* User Sheet remains posted by each computer.

■ Work through the following sections of the *Geo-Logo* Teacher Tutorial:

Making Shapes That Follow Rules   127

Polygons with Moves and Turns   129

How to Choose a New Activity; How to Draw with Moves and Turns; How to Use the Repeat Command; How to Make Polygons with Moves and Turns; How to Use the Turtle Turner and the Ruler; How to Begin Drawing at a Different Location; How to Draw in Color; Comparing Moves and Turns to Coordinate Commands

Angle and Turn Game   134

■ Plan how to manage the computer activities, depending on computer availability.

With five to eight computers: Follow the investigation as written. Half the students work at the computer in pairs or threes while the other half work off computer. Groups then switch.

With a computer laboratory: Begin Session 4 off the computer with everyone working with Power Polygons. When students have finished, move to the computer lab for the on-computer activities.

Do the reverse for Sessions 6 and 7: Begin in the computer lab with a demonstration, followed by student work on the computer. When students have finished, they can work on the off-computer assessment activity.

Session 8 is off the computer, with everyone working with Power Polygons.

For Session 9, move to the computer lab for the computer game. When students have finished, they do the off-computer activities.

With fewer than five computers: Introduce the on-computer activity at the beginning of Session 4 and immediately assign some students to begin cycling through it. Make and post a schedule, assigning about 20 minutes of computer time for each pair of students throughout the day.

As students finish this activity, introduce the Session 6 computer work and again assign students to cycle through it.

Introduce the Session 9 on-computer activity at the beginning of Session 8, so that students start cycling through the on-computer activity while the rest are starting the Session 8 off-computer activities. Students may have to complete their computer work for this investigation while you begin a new investigation or engage the rest of the class in a Ten-Minute Math activity.

Sorting Polygons

What Happens

After a brief homework review, students begin sorting polygons: the three-sided figures and four-sided figures from the Guess My Rule Cards. They think of different ways to categorize triangles and quadrilaterals and then complete the statements "All triangles...." "Some triangles...." "All quadrilaterals....," and "Some quadrilaterals...." Finally they play Guess My Rule with all the cards. Student work focuses on:

- exploring attributes of triangles and quadrilaterals
- sorting and classifying triangles and quadrilaterals
- developing vocabulary to describe special triangles and quadrilaterals
- developing an understanding of parallel lines

Materials

- Guess My Rule Cards (1 deck per pair, and 1 deck per student, homework)
- Rubber bands or paper clips (1 per pair)
- Transparent Guess My Rule Cards (1 set)
- Overhead projector
- Loop of string or blank transparency
- Chart paper
- Power Polygons (available)
- Student Sheet 6 (1 per student, homework)
- Student Sheet 7 (1 per student, homework)
- Student Sheet 8 (1 per student, homework)

Activity

Homework Review

Take a few minutes to review and share students' work on the class chart where they have been listing words or examples related to the different polygons.

When you have talked about all the words students have found, suggest a few others, maybe for a prefix for which students have found no examples. Challenge students to guess the meanings. For example:

A 10-sided polygon is a *decagon.* **What do you think a decaliter is?** (a metric measure that is equal to 10 liters) **What do you think a** *decathlon* **might be?** (an athletic contest with 10 events)

Identifying Triangles by Their Angles

Distribute a prepared deck of Guess My Rule Cards to each pair. If students label the back of each card with their initials, decks that get mixed together can be easily restored.

Ask students to sort through their deck and pull out all the *three-sided* figures. They can store the rest of the cards in the envelope for now.

When we talked about the names of polygons, we said that one way we could name any polygon is by how many sides it has. Look at all the three-sided polygons you have in front of you.

What do we call three-sided polygons? Are all of these three-sided figures triangles? Why or why not?

Many students think that only the more familiar triangle shapes are real triangles. The **Teacher Note,** The Visual Side of Learning (p. 41), explains how students can misunderstand verbal definitions of polygons.

In this discussion, students may not reach agreement that all the three-sided shapes are triangles. It's OK to leave the issue unresolved, as students will be writing about the question for homework tonight. The **Dialogue Box,** Are All Three-Sided Polygons Triangles? (p. 44), shows how a follow-up discussion, after the homework, unfolded in one class.

Types of Angles Using the transparent Guess My Rule Cards, choose any two of the four right triangles (1, 4, 5, 8). Place them on the overhead so that the right angles are not oriented the same way. Either draw a circle around them on a blank transparency or enclose them with a loop of string. Put a triangle *without* a right angle just outside the circle or loop. Scatter the other triangles in a pool.

I have picked out two triangles that have something in common and put them inside the loop. One triangle that doesn't go with them is on the outside. What other triangles from our collection could be placed in the loop with these two?

Many students will be able to choose the other two right triangles from the pool and give reasons for choosing them. Some may say that all these triangles have square corners, or 90° angles. However, most students will have had little experience talking about and identifying angles, so take a few minutes to explore them further.

Start by asking students to stand up and hold both arms straight out in front of them, then open one of their arms to the side to form a right angle.

You have just made a right angle, or a 90° angle. Here is another way to think about angles: Angles are *inside* shapes. Imagine that your two arms form two sides of a shape. That makes them two sides of an angle. Your head is at the vertex, or the corner, of the angle. The measure of the angle, 90°, tells how much your arms are open.

Ask students to show with their arms an angle smaller than 90° and then one larger than 90°. Explain that any angles that are not right angles are either *acute* (smaller than 90°) or *obtuse* (larger than 90°). Ask students to find triangles in the Guess My Rule Cards that show these three kinds of angles. By now, they should recognize that your category in the loop is triangles that have right angles.

Next introduce the idea of *sorting* triangles by their angles. You might introduce the names of some of these triangles, but don't overemphasize the vocabulary; the characteristics of the shapes are more important than the terminology.

Looking only at their angles, we can sort triangles into three types: right triangles, obtuse triangles, and acute triangles. (Acute triangles are more typically named by the number of sides that are the same length.) See the **Teacher Note,** Classification of Triangles and Quadrilaterals (p. 42), for more detail on this.

Now is a good time to talk about and show how to use a square from the Power Polygon set as a "right-angle tester" to see if an angle is right, acute, or obtuse. Students who worked in the grade 4 unit, *Sunken Ships and Grid Patterns,* may remember that a right-angle tester can be made from any scrap of paper by folding it in half first one way, and then the other, creating a reliable 90° angle.

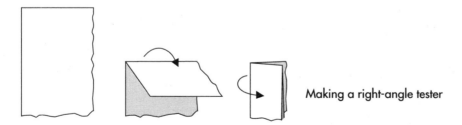

Making a right-angle tester

You might also show students the convention for labeling a right angle with a little square.

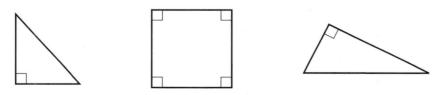

Sorting Triangles

Using the twelve triangles from the Guess My Rule Cards, student pairs or groups of four categorize triangles in various ways as they play Guess My Rule. One student silently thinks of a category and picks two triangle cards that fit in that category. He or she then challenges the others to find more shapes that belong in the group and to guess the category.

Students use notebook paper to record the categories they find and the numbers of the triangles that fit them.

When all the groups have found two or three categories, bring the class together and make a list of their ideas on chart paper with numbers (and maybe sketches) of the shapes in each category. Write the descriptions as students say them and suggest formal names if you like, remembering that the *concepts* are more important than the vocabulary.

Students might describe categories like those shown in the chart completed by one fifth grade class (below).

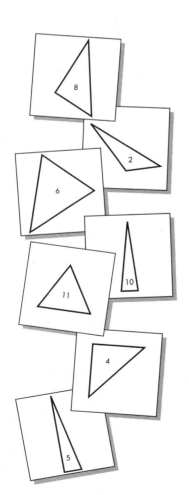

Skinny triangles 5, 10
long sides 9, 2, 5
▶ 90° angles (right triangles) 1, 4, 8
sides aren't the same length 2, 7, 5 ◀
all even/same sides] 11, 12
(equilateral)
not perfect /all different lengths 2, 5, 7 ◀
(scalene)
spiky 5, 10, 9
stretched triangles (obtuse) 2, 7, 9
(If you have another like it, they would
fit together to make a rectangle 1, 8, 4
(right triangle)

After this activity, students return their triangle cards to the Guess My Rule decks, using rubber bands or paper clips to keep them separate from the other cards. You might end Session 1 at this point.

Toward the end of Session 1, students might begin working on their home-work (see p. 40), writing about whether all three-sided polygons are triangles and why.

All Triangles, Some Triangles

On chart paper, set up lists headed *All triangles...* and *Some triangles....* Display on the overhead the twelve transparent Guess My Rule Cards that are triangles.

You have told me that all triangles have three sides. *[Record this fact under "All triangles...."]* **What else do all triangles have?**

You also noticed there are some differences among triangles. Look at these triangles and at the chart of categories you came up with for sorting them. How can we complete this statement: "Some triangles have..."?

Record students' suggestions on the list. To get them started, ask:

What is true of the *sides* of some triangles? What *kinds of angles* do some triangles have?

As students make suggestions, they give numbers of triangles on the Guess My Rule Cards that fit their description. If they can find no triangles that fit a certain description, they may draw an example.

List the different attributes students mention. Ask students if they know the name for each kind of triangle they are describing. Record the name on your list as they tell you, or, if no one knows, suggest it yourself. (The **Teacher Note**, Classification of Triangles and Quadrilaterals, p. 42, gives the names and categories of triangles and quadrilaterals.)

❖ **Tip for the Linguistically Diverse Classroom** Include sketches and symbols on the list of attributes to ensure that all students understand the categories.

All triangles...

have 3 angles, corners, points

have 3 sides or lines that connect

have 1 middle, are closed, are polygons

have 3 angles that add to 180°

Some triangles...

have a 90° angle (right triangle)

have 3 different side lengths (scalene)

have all sides and angles the same (equilateral)

have 2 sides the same length (isosceles)

are different shapes—short, fat, tall, skinny

have all small angles (acute angle, less than 90°)

have one big angle (obtuse angle, bigger than 90°)

have a right angle and 2 equal sides (isosceles right or right isosceles)

Some students may say that all triangles have 180° inside. If students suggest this, ask them to explain how they know or to show you what they mean. If they can't, you might demonstrate this way: Start with any paper triangle, tear off the three angles, and line them up to prove they form a straight line, or 180°. You can use the same method later to show that the angles of a quadrilateral add up to 360°.

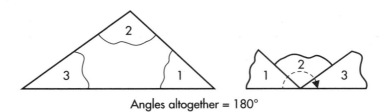

Angles altogether = 180°

When students have run out of ideas for attributes of triangles, post the list and tell them you will continue to add to the list throughout the unit as they come up with more ideas.

Identifying Quadrilaterals

Student pairs take out their decks of Guess My Rule Cards and sort out all the four-sided shapes. (There are 15 four-sided polygons in the deck, shapes 13 through 27.) Ask if there is a single name we can use for all these shapes. As necessary, remind students of the chart the class made in Session 1 that lists the names of polygons by number of sides.

Some students may think the name for four-sided shapes should be *squares or rectangles*. Encourage them to look for four-sided shapes that are not squares or rectangles so they can be convinced that other quadrilaterals exist and that squares, rectangles, and all other four-sided shapes are quadrilaterals, just as all three-sided shapes are triangles.

Parallel Sides Place a loop of string on the overhead screen or display the transparency with a circle on it. Then place inside the loop three or four of the quadrilaterals with at least one pair of parallel sides (13, 16, 18, 20, and 21–27). As you place them, intentionally orient the parallel sides in different ways. Leave the other quadrilaterals scattered in a pool.

The quadrilaterals I have put inside the loop have something in common. Instead of guessing what the rule is, name the number of another shape that you think might fit my rule. If I place that shape in the loop, it fits the rule. If I put it outside the loop, it does not fit the rule. Keep track with your own cards at your desks. Make a group of shapes that fit and set aside those that do not.

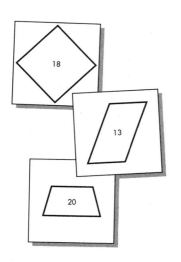

As students suggest shapes, stop them from guessing the rule out loud until it seems that many of them know it. When students finally state the rule, ask them to tell you which sides in these shapes are parallel, and to describe what *parallel* means.

Parallelism is a notoriously difficult concept for children. Many students will have difficulty seeing that two lines can be parallel when they are not directly across from each other. Through discussion, try to generate a classroom definition. You might suggest that students think of *parallel* as meaning "going in the same direction," or compare parallel lines to train tracks.

Sketch on the board some examples of figures with parallel lines, as well as some with no parallel lines. Ask students to point out the lines that are parallel or to demonstrate that the lines are parallel or not.

Parallel lines   No parallel lines

Using the set of quadrilaterals on the Guess My Rule Cards, students now find other ways to categorize them (just as they did earlier with triangles). They record their categories on notebook paper, along with the numbers of the shapes that fit each category.

Sorting Quadrilaterals

Again, record students' categories on chart paper, with numbers and sketches of the shapes that fit each category.

To encourage students to compare categories, you might ask them what shapes appear in many categories (shapes 18 and 21), and why that is so. This is a chance for students to discuss such questions as whether all squares are rectangles and all rectangles squares. The **Dialogue Box, Are Squares Rectangles? (p. 45),** illustrates one class discussion of this question.

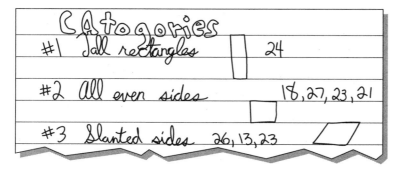

Categories
#1 all rectangles   24
#2 all even sides   18, 27, 23, 21
#3 Slanted sides 26, 13, 23

Attributes of Quadrilaterals

Just as you did with triangles, set up two lists on chart paper headed *All quadrilaterals...* and *Some quadrilaterals....*

How many sides in a quadrilateral? Do *all* quadrilaterals have four sides? *[Record this fact under "All quadrilaterals...."]* **What else do all quadrilaterals have to have?**

What are some of the differences among quadrilaterals? Look at the shapes, and think about the categories you came up with for sorting them. How can we complete the statement, "Some quadrilaterals have..."?

What is true of the sides of some quadrilaterals? What kinds of angles do some quadrilaterals have?

List the attributes students mention. Ask them to give the numbers of quadrilaterals that have the attribute they suggest, or draw one on the board. If they can name the shape specifically, record the name with the related attribute.

If students say a quadrilateral has 360° inside, ask them to explain how they know or to show what they mean. If they can't, demonstrate as you did for triangles, by tearing angles from paper quadrilaterals and arranging them to prove they form a full circle, or 360°.

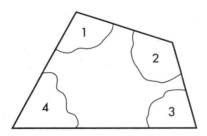 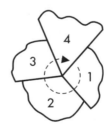

Angles altogether = 360°

All quadrilaterals...	**Some quadrilaterals...**
have 4 lines, segments, or sides	have a right (or 90°) angle or turn (are perpendicular)
have no rounded corners, no crossing lines (are polygons)	have parallel lines (going in the same direction, equidistant from each other)
have angles that add up to 360°	have all equal sides (square or rhombus)
can have different shapes and sizes (don't have to be perfect)	have 2 sides parallel and other 2 sides parallel (parallelogram)
have 4 angles, corners, points, or vertices	have no sides the same (nonregular)
	have acute angles and obtuse angles

Continue to add to this list as students come up with more ideas about quadrilaterals. Be sure students understand they are not restricted to thinking about the shapes included in the Guess My Rule Cards. Encourage them to find and draw examples of quadrilaterals (and also triangles) that are not represented on the cards. Can they find the names for any of these special quadrilaterals and triangles?

Playing Guess My Rule

Students play Guess My Rule in groups of three or four, using a full deck of Guess My Rule Cards. For the rules they present to each other, they try to find attributes of triangles and quadrilaterals beyond those that have been listed in class, and attributes of other polygons pictured on the cards.

To start, one student (or pair) is the leader. The leader decides on a rule, writes it on a piece of paper, and keeps it hidden.

❖ **Tip for the Linguistically Diverse Classroom** When students who are less proficient in English are the leaders, they can illustrate their rule with a drawing rather than writing it out. They might label their drawing with numbers, symbols (such as the equals sign), and other designations.

Each group uses a piece of paper as a "loop" for categorizing. The leader indicates a few shapes that fit the rule and a few that don't, placing those that fit on the paper, those that don't to one side (off the paper). Without guessing the rule aloud, other players take a new shape and place it on or off the paper to show whether they think it follows the rule. The leader tells whether that placement is correct or not.

No one may guess the rule aloud until everyone has had a chance to try out several shapes, or until there are no shapes left. After one round, the group plays again with a new leader.

As they play, groups list the categories they have used. If students would like more shapes to play with, they might take one of each shape from the Power Polygons.

Circulate to be sure everyone understands the game and that all students are participating.

Follow-Up Discussion Toward the end of Session 3, call the class together to share some of the new categories students used in Guess My Rule, showing examples of each. This is another chance for you to introduce precise vocabulary and to add to the lists of *Some triangles*.... and *Some quadrilaterals*....

Sessions 1, 2, and 3 Follow-Up

 **Homework**

Is Every Three-Sided Polygon a Triangle? After Session 1, give each student a copy of Student Sheet 6, Is Every Three-Sided Polygon a Triangle? Students write about whether or not they think all three-sided polygons are triangles and give reasons for their thinking. Be sure they also provide sketches to illustrate their ideas. Plan to follow up with a class discussion of what they have written.

❖ **Tip for the Linguistically Diverse Classroom** Students may write their answers in their primary language or indicate their ideas with mostly drawings.

Is Every Square a Rectangle? Is Every Rectangle a Square? After Session 2, give each student a copy of Student Sheet 7, Is Every Square a Rectangle? Is Every Rectangle a Square? Students answer these questions and give reasons for their thinking. They also draw sketches to illustrate their ideas.

Use this homework to evaluate students' understanding of squares and rectangles. In particular, look for the following:

1. Do students know what is required for a shape to be a rectangle and what is required for it to be a square?

2. Do students recognize that the requirements for a square are more restrictive?

3. Can students explain that all squares are rectangles, but not all rectangles are squares?

At this time, students should be clear about item 1. If some are not, take time to work with them on identifying and describing squares and rectangles. Students will improve their understanding of items 2 and 3 in the next few sessions.

How to Play Guess My Rule with Shapes After Session 3, students play Guess My Rule at home with family or friends. They each take home a copy of the two sheets of Guess My Rule Cards and a copy of Student Sheet 8, How to Play Guess My Rule with Shapes. If students are unlikely to have scissors at home, give them time in school to cut the cards apart. As they play, they should keep track of any new categories that might help them write new "some" statements about triangles and quadrilaterals.

The Visual Side of Learning

Contrary to what many people believe, students frequently do not use definitions of concepts in their thinking. Instead, they use concept images: a combination of all the mental pictures and properties that they have associated with a concept.

Illustrations used in instruction have a strong influence on these concept images. For example, if students see figures only in certain "standard" positions, they may assume that the figures occur only in those positions (as illustrated by the common misconception about squares, discussed below).

Even when students learn correct, standard verbal descriptions or definitions of a concept, their concept images, influenced by their visual experience, tend to rule their thinking. Therefore, they need to combine each verbal description they encounter with a wide variety of visual examples.

It is important to provide the variety of visual examples both when a concept is introduced and later, to help students construct meaningful verbal descriptions and definitions from the examples. If verbal descriptions are to be of real use, students need to understand them thoroughly. That is why the activities in these investigations are set up so that students construct their definitions based on many different images.

For example, many students may start the unit with limited ideas about triangles. You can be instrumental in encouraging them to understand which characteristics are relevant to the definition of a triangle (three sides, closed figure), and which characteristics are irrelevant (size and orientation, for example), and to synthesize the resulting visual images and verbal definition.

In one classroom, a teacher drew two squares on the overhead, oriented like these.

When asked if both were squares, several students said no—that one was a diamond. After some discussion, the teacher pointed out that nothing in the rules for squares says that they must have horizontal and vertical sides.

The teacher then drew two right angles, oriented as below.

After students established that their right-angle tester (a plastic square) worked for both angles, the teacher drew a third right angle:

This time, after a little discussion, a student volunteered that "nothing in the rules for right angles says they have to sit flat." The teacher acknowledged this application of the definition.

Discussions like this help students to visualize more possibilities for squares and right angles. They also help them consider and apply the definitions of the terms, recognizing that what the definitions do *not* say can be as important as what they do say.

Classification of Triangles and Quadrilaterals

Classification systems help us to organize the world into categories that often are hierarchical and overlapping. For example, a person might live in a particular town, which is in a particular state, which is in the United States, which is in North America, which is in the Western Hemisphere. If I say that I live in that town, you know that I live in all the other places that include the town; for example, everyone who lives in Cleveland also lives in Ohio, and so on. In a hierarchical classification system, we can't make the same kind of assumptions in reverse; it is not true that all people who live in Ohio live in Cleveland.

We use a hierarchical classification system to sort geometric figures. The activities in this unit help clarify the classification of triangles and quadrilaterals.

Triangles

There are two ways to classify triangles, by their angles and by their sides. Classified by their *angles,* triangles are right (one 90° angle), acute (all angles smaller than 90°), and obtuse (one angle greater than 90°). These categories are illustrated by the horizontal loops in the diagram. Classified by their *sides,* triangles are scalene (no sides the same length), isosceles (at least two sides the same length), or equilateral (all sides the same length).

Consider two triangles:

Triangle A has two sides the same length and a 90° angle, so it is both right and isosceles: a right isosceles triangle. Such double classifications take some time to get used to.

There are also more complex relationships: Triangle B is acute because it has three acute angles; it is equilateral because it has three equal sides; and that means it is also isosceles, because at least two of its sides are equal. That is why it lies inside three loops. So Triangle B is an *equilateral* triangle, all of which are members of the isosceles triangle family, which are members of the triangle family, which are members of the polygon family.

We usually speak of an object by its most restrictive category: If a triangle has three equal sides, we call it *equilateral* rather than isosceles or just a triangle. In providing illustrations, we do the opposite: We provide the least restrictive example. To illustrate an isosceles triangle, we draw one with two equal sides, not three; for a general triangle, we draw a scalene triangle without a right angle.

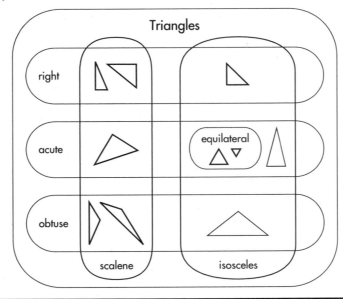

Continued on next page

Quadrilaterals

There are two ways to classify quadrilaterals. The first asks if they have parallel sides. *Trapezoids* have at least one pair of parallel sides. *Parallelograms* have two pairs of parallel sides.

The second way of classifying quadrilaterals concerns the lengths of their sides. *Kites* and *chevrons* have two pairs of equal adjacent sides. *Isosceles trapezoids* have one pair of equal opposite sides. *Parallelograms* have two pairs of equal opposite sides. *Rhombuses* (or rhombi) are members of the parallelogram family that have all four sides equal.

The angles are what make rectangles special. Rectangles are members of the parallelogram family with four equal angles. Squares, then, are in many families, including rectangle, rhombus, and parallelogram.

The diagram below shows the complicated hierarchical classifications of quadrilaterals, which include trapezoids and parallelograms.

Trapezoids can be further classified into isosceles (one pair of opposite sides equal) or not isosceles, and right (one right angle) or not. Isosceles concave quadrilaterals are commonly known as chevrons. Kites are another special kind of quadrilateral.

Such traditional classifications are just one useful way to sort geometric figures. We could just as well declare that rectangles *cannot* have all equal sides, and then squares would not be in the family of rectangles. Students often prefer this partitioning way of classifying. Only with time will they come to see the advantages of hierarchical classification—for example, economical definitions and logical inference (if you know a square is a rectangle, you know it has all the properties of rectangles).

At this age, students will benefit from thinking and communicating about the properties of polygons, but they need not have the whole classification system in mind.

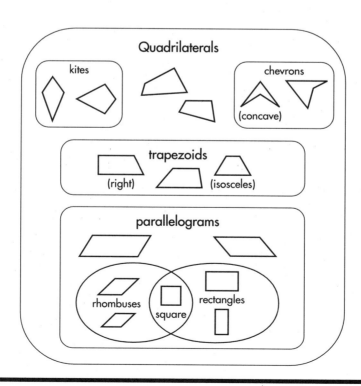

Are All Three-Sided Polygons Triangles?

As this class discusses categories for the triangles on the Guess My Rule Cards, some students are confused about whether *all* three-sided polygons should be called triangles.

We're looking for categories for sorting three-sided polygons. So far we have shapes with two equal sides, shapes with all equal sides, shapes with all unequal sides, and shapes that if you had another shape just like it, you could make a square or rectangle. What else?

Robby: How about shapes that look like triangles. Examples are 3, 11, 6, 8, and 12. Not 9.

What must a figure have to look like a triangle?

Robby: Three equal sides. Like one would be left side 15 cm, right side 15 cm, and the bottom connecting the two 15 cm.

Robby feels the other three-sided polygons aren't triangles. What do the rest of you think? Does a polygon have to have three equal sides to be called a triangle?

Yu-Wei: The shape of 9 is a stretched triangle, so it's not a triangle.

Julie: I disagree, because other ones that are sort of stretched like 6 and 8 *are* triangles.

Robby, which ones do you feel are not triangles?

Robby: I think 7, 10, 9, 2, and 5 aren't.

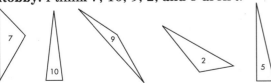

[There is a lot of discussion at this point, and the teacher asks to hear more about what makes a shape a triangle.]

Help! What *is* a definition of a triangle?

Lindsay: It has to have three angles. You know that because *tri*-angle; *tri* means "three."

If *triangle* means three angles, where do the sides fit in?

Matt: Anything with three sides is a triangle.

Manuel: I agree with three sides, but I think they all have to be equal.

Leon: Then how come they have triangles called isosceles and scalene?

[Class ends with the question unresolved. Discussion continues the next day.]

We ended math yesterday with a big question: Are all three-sided polygons triangles? What did you decide?

Cara: I wrote: "I think a triangle is a shape that has three sides and three corners. I think that triangles don't have to have the same length in the sides."

Robby: I made a category called "Triangles," and I wrote, "A triangle should have even sides and three corners."

So what about three-sided shapes that don't have even sides? What would you call those?

Robby: I don't know.

Antonio: I used to think like Robby, but now I think any three-sided polygon is a triangle. I didn't know what to call those other shapes except triangles.

Julie: I agree. I think a triangle's something with three sides no matter how long they are.

Manuel: Triangles can have different sides, and then have other names that mean they're not equal, like *right* or *isosceles* triangles.

It's important that we're coming to agreement about these terms as a class.

Are Squares Rectangles?

This class is discussing quadrilaterals when a controversy arises: Is a square a rectangle? Is a rectangle a square? The teacher chooses one shape, a square, on which to center the discussion.

What is this *[holding up a square]*: **a square, a rectangle, or both?**

Yu-Wei: It's a square.

Becky: But a square is a rectangle, isn't it?

Amy Lynn: No, a rectangle is a square, but a square isn't a rectangle.

What do we know has to be true about a square?

Amy Lynn: All 90° angles and even amounts of edges or sides.

So what about this shape? *[The teacher draws a square on the board.]*

Becky: You know what you could call it? An equilateral rectangle.

Becky, can you explain your reasoning? I see a lot of people don't agree.

Becky: A rectangle is four sides that have parallel sides somewhere else in the shape. A square has those things, but instead of two sides being different, they're all the same.

OK, for a square, the sides all have to be equal. Now, could we call a square a rectangle?

Desiree: A rectangle doesn't have all sides equal; it only has two.

Julie: A rectangle has two equal sides. It has to have two equal sides, and a square has four equal sides. But in those four equal sides, a square has the two it needs to be a rectangle.

Amy Lynn: Well, then you should be able to reverse that too.

Julie: But a rectangle can't be a square, because a rectangle has two sides equal and then another two sides equal but different from the other and a square *has* to have four equal sides. Becky's right, a square can be a rectangle.

Amy Lynn: I agree, but a rectangle can be a square because it's just the opposite.

Manuel: A square can always be a rectangle, but only a rectangle with four equal sides can be a square.

Amy Lynn: But then you'd just call it a square.

Manuel: You could call it a square or a rectangle.

Amir: It's like first and last names in families. You can call them square rectangles.

Lindsay: Or an equilateral rectangle, like Becky said.

Making Shapes That Follow Rules

Materials

- Power Polygons (1 bucket per 6–8 students)
- Student Sheet 3 (available in class; 2 per student, homework)
- Student Sheet 9 (1 per student)
- Student Sheet 10 (1 per student)
- Student Sheet 11 (1 per student, homework)
- Transparencies of Student Sheets 9 and 10
- Overhead projector
- Computers

What Happens

Using Power Polygons off the computer and *Geo-Logo*'s coordinate commands jumpto and setxy on the computer, students make triangles and quadrilaterals that fit given descriptions. They then discuss which of the shapes are impossible to make and which are difficult to make. Students' work focuses on:

- generating geometric figures from descriptions of their properties
- describing geometric figures orally and in writing
- using *Geo-Logo* commands setxy and jumpto and the Label Lengths tool to draw geometric figures with specific properties
- sorting and classifying triangles and quadrilaterals

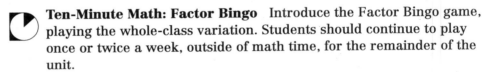 **Ten-Minute Math: Factor Bingo** Introduce the Factor Bingo game, playing the whole-class variation. Students should continue to play once or twice a week, outside of math time, for the remainder of the unit.

Each player will need a Multiplication Table and a crayon or marker. Calculators are optional. When playing in small groups, each group will need a deck of Factor Bingo cards. Players take turns turning over a card and calling the number on it.

When a number is called, each player marks a factor of that number on his or her own Multiplication Table. (They write the number that was called in each marked square, to check their accuracy later.) Thus, if someone turns over a 100 card, any factor of 100 on the Multiplication Table—1, 2, 4, 5, 10, 20, 25, 50, or 100—can be marked.

The player who turns over a Wild Card decides on the number to be used. The game continues until a player marks five squares in a row for a Bingo.

For full direction and variations on Multiple and Factor Bingo, see p. 112.

At the beginning of Session 4, collect any additional "Some triangles..." and "Some quadrilaterals..." statements that students may have found while playing Guess My Rule at home. Again, supply formal names to fit their descriptions, and write those terms on the list for students to use as a reference. Encourage students to add to the lists throughout the investigation as they find new kinds of triangles and quadrilaterals.

Adding More Attributes

Distribute the Power Polygons so that groups of six to eight students have a bucket to share. Students work with partners within their groups. Introduce the activity students will be doing during the next two sessions, both on and off the computer.

We're going to play a game that's a little like Simon Says, in which you have to follow certain rules. Here's how it works: Each of you may take any of the Power Polygon shapes—you can put more than one shape together if you want—to end up with one shape that follows all the rules I give you. Here are the rules your shape must follow:

   1. Its outline is a five-sided polygon.

   2. It has at least one right angle.

   3. It has exactly one pair of parallel sides.

Write the three characteristics on the board so students can refer to them. If students need a reminder of what *parallel* means, invite volunteers to come to the board and draw a shape with parallel sides.

When you've made a shape that fits all three rules, check to see that your partner agrees.

When students have made shapes and checked with their partners to see if the shape follows all the rules, several students share their solutions (possibly on the overhead). If there is disagreement about whether a shape follows the rules or not, students from each side explain their reasoning until they reach agreement.

For this introduction to the activity, be sure to include the idea of parallel sides and right angles. See the **Dialogue Box,** What Rules Are You Following? (p. 54), for one class's discussion of their solutions.

Following the Rules

Rules
- 5 sides
- at least 1 right angle
- exactly 1 pair of parallel sides

Can You Make These?

Give each student a copy of Student Sheet 9, Can You Make These Triangles? and Student Sheet 10, Can You Make These Quadrilaterals?

These sheets give you the rules for ten rounds of the game we just played. Today and tomorrow, you will work in groups of four to make triangles and quadrilaterals that follow certain rules. For every figure, you will try it two ways: on the computer, using the `setxy` and `jumpto` commands, and off the computer, using the Power Polygons.

Record both your on-computer solution and your off-computer solution on these sheets.

One warning: Some of the figures may be impossible to make if you really follow all the rules. If one or two of you decide that a figure is impossible, the rest of your group must also try it and agree with you that it can't be made.

Place the transparencies of Student Sheets 9 and 10 on the overhead to go over how the sheets are organized and how to record solutions on them.

Name _____   Date _____

© Dale Seymour Publications®

Student Sheet 9

Can You Make These Triangles?

	Length of sides	Angles	Name of shape	Setxy points that make this triangle	Sketch of Power Polygons that make this triangle
tri1	your choice	1 right angle			
tri2	your choice	2 right angles			
tri3	3 equal sides	3 equal angles			
tri4	your choice	3 angles smaller than a right angle			
tri5	0 equal sides	1 angle larger than a right angle			

Do you think any of these triangles are impossible?
If so, pick one and write about how you could prove it is impossible.

Investigation 2 • Sessions 4–5
Picturing Polygons

Point out that the first column simply numbers the different triangles or quadrilaterals students are to make. The next two columns on Student Sheet 9 (three columns on Student Sheet 10) are the rules students are to follow when making the figures. For example, tri5 has no equal sides and one angle larger than 90°.

In the *Name of shape* column, students write the specific name, if they know it, for the shape they make, or they write "impossible." In the next column, students list the setxy points they used on the computer, recording them either as *Geo-Logo* commands [30 20] or as ordered pairs (30, 20). In the last column, students sketch the Power Polygon pieces they used. Students should label each shape with its Power Polygon letter so they can be sure later which ones they used. Students will finish the student sheet for homework.

Students will once again be working in the Polygons with Coordinates activity. Before they start the computer activity, introduce another *Geo-Logo* tool, Label Lengths.

On Computer

Making Shapes with *Geo-Logo*

As you try to make the triangles and quadrilaterals on the computer, you will sometimes need to prove things—like your angles are 90°, or more or less than 90°, or your sides are the same length. There's a tool on the computer's tool bar that can help you check the lengths of your shape's sides: the Label Lengths tool.

Refer students to the icon for this tool on the *Geo-Logo* User Sheet.

Label Lengths

When you click on the Label Lengths tool, it will label all the sides of your figure with their length in turtle steps. In these activities, the turtle's step is set at 1 mm.

You might also want to bring a right-angle tester—such as Power Polygon square B, or a scrap of folded paper—with you to the computer to help you check the angles in your figures.

A "turtle step" is the distance the turtle travels given the command fd 1. In these activities, the length of a turtle step is usually set at 1 mm; however, this can be changed. See the discussion of **Scale Distance** in the *Geo-Logo* Tutorial, More About *Geo-Logo*, in the section on Menus (p. 148).

Working on the Computer Students write procedures, using *Geo-Logo's* coordinate commands, that draw examples of shapes that fit the rules on Student Sheets 9 and 10. Make available copies of Student Sheet 3 (Coordinate Grids) for students to use in planning their shapes.

Off Computer

**Making Shapes
with Power
Polygons**

Students use Power Polygons to find or construct examples that fit the rules on Student Sheets 9 and 10. They also use off-computer time to plan coordinates for the shapes they will make on the computer, working with the coordinate grids on Student Sheet 3.

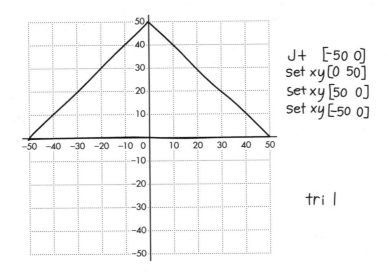

```
J+  [-50 0]
set xy[0 50]
set xy [50 0]
set xy [-50 0]


tri I
```

Classroom Management

Students work in groups of four. Two of the four spend the remainder of Session 4 on the computer, while the other pair work at their desks. When it is time to switch at the beginning of Session 5, the group confers to see which of the specified shapes have been made successfully. They then work on the other shapes for the rest of the work time (the first half of Session 5).

Encourage all students to try making the equilateral triangle and rhombus on the computer using setxy and jumpto, and to think about whether they are truly impossible or just very difficult.

For both the on- and off-computer groups, emphasize that the goal is not finishing all ten shapes, but thinking and learning about the characteristics of each polygon. If students finish both sheets, they can look for more than one solution for some polygons. Remind students that some shapes are indeed impossible.

Observing the Students   Circulate among the students at the computer to make sure they understand how to use *Geo-Logo* for this activity. Ask them to prove that the sides in their tri3, quad1, or quad2 are equal. They can use the Label Lengths tool for this purpose. As you observe, look for the following:

■ How are students organizing the task?

■ Are they able to assimilate more than one characteristic in picturing a shape?

■ How are students deciding which shapes are impossible? Can they prove it? Can they distinguish shapes that are impossible from shapes that exist but are difficult to make accurately on the computer with setxy?

■ Can they think about how the Power Polygons might help them picture a particular problem? Can they use a particular shape and its characteristics to help them find a solution?

The shape quad2 will be difficult for students if they try to make two sides of the shape run along parallel grid lines. You might suggest that they take a diamond (rhombus) from the Power Polygons and orient it in different ways against the computer screen to see if they can plan where the coordinates would be.

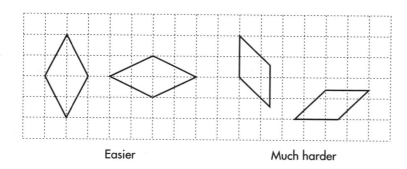

Easier   Much harder

About halfway through Session 5, call students together to discuss what they have found and to decide on the names for the shapes.

Discussing the Shapes

Triangles   Display the transparency of Student Sheet 9 and go through the triangles one by one. As a few students share different triangles they made for the same set of rules, fill in your transparency.

What strategies did you use to make the triangle? Were you able to make it with both setxy and the Power Polygons?

When you get to tri2 (two right angles) and some students observe that it is impossible to make, ask them to explain why. Also focus on tri3, an equilateral triangle.

Which triangle was particularly difficult to make? Why was it difficult?

Students may suggest that it is possible to make tri3 with the Power Polygons but impossible on the computer. Ask if anyone was able to make a triangle with three equal sides on the computer. If so, how could they prove their triangle is equilateral?

Acknowledge that it is very difficult to make a triangle with three equal sides using only `setxy` commands. In the next sessions, students will learn other commands that make the task easier.

If some students believe that tri2 is possible, encourage them to explain why and tell them they will have another opportunity to try to make it on the computer in the next session, with different computer commands. The **Dialogue Box,** Which Are Impossible? (p. 55), presents one discussion of this problem.

Quadrilaterals Display the transparency of Student Sheet 10 and, just as you did for Student Sheet 9, collect solutions from students to fill in the chart. Encourage students to show different quadrilaterals they made for the same set of rules. Ask about their strategies for making them.

Ask students who believe that quad3 is possible how they would make it. Also ask which shapes were particularly difficult to make. As students discuss shapes that are possible but difficult to make accurately with `setxy`, encourage them to use the names for the shapes, to state their attributes, and to sketch them or show them with Power Polygons.

In the next math class, we will review *Geo-Logo* **commands such as** `fd` **(forward) and** `rt` **(right turn), and you will have a chance to try making tri3 and some of the other shapes with these move and turn commands.**

Sessions 4 and 5 Follow-Up

Coordinate Grids and "Quad" or "Tri" Commands   After Session 4, students take home copies of Student Sheet 3, Coordinate Grids, along with their partially filled in Student Sheets 9 and 10, and write setxy commands for any "quad" or "tri" shapes their group has not yet made on the computer. They determine which shapes are impossible and write a few sentences about why they are impossible. Students should also write about why certain shapes are very difficult but not impossible. Remember to save Student Sheets 9 and 10 for use in Sessions 6 and 7.

❖ **Tip for the Linguistically Diverse Classroom**   Students may respond to the writing part of this assignment in their primary language. Alternatively, they might try visual diagrams, copying from Student Sheets 9 and 10 the part or parts of each set of criteria that make it impossible to draw the shape.

Find the Fourth Vertex   After Session 5, students take home Student Sheet 11, Find the Fourth Vertex. As necessary, review the properties of a parallelogram. Don't tell students that there are three possible solutions: (10, 50), (50, –30), and (–10, –30).

Students might also write all the coordinates to make a different parallelogram, being sure to use some negative coordinates.

Writing *Geo-Logo* Procedures   If students are familiar with *Geo-Logo* commands fd (forward) and rt (right turn), they write *Geo-Logo* procedures to draw a square and a rectangle that is not a square, in preparation for learning the repeat command in Sessions 6 and 7.

 **Homework**

 Extension

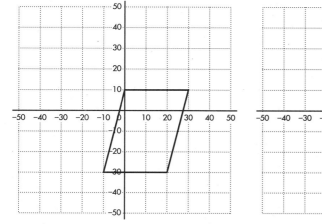 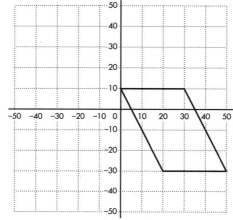 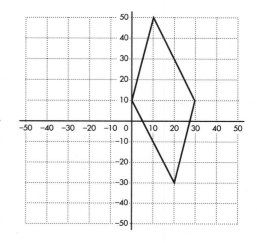

Find the Fourth Vertex: Solutions

What Rules Are You Following?

These students are discussing the polygons they created to fit three specific rules in the activity Following the Rules (p. 47). One students' proposed shape doesn't fit her classmates' image of the set of rules.

Mei-Ling's figure

Our rules were that the shape is a polygon; it has five sides; it has at least one right angle; and it has exactly one pair of parallel sides. Does Mei-Ling's figure follow our rules?

Kevin: No, Mei-Ling's isn't a house shape at all.

Manuel: Well, it does have five sides and a right angle… and one pair of parallel sides.

Jeff: I guess so. But it doesn't poke out.

Kevin: Yeah. It's got a turned-in angle, but…it does follow the rules. I just thought it couldn't be turned in.

Sometimes we call a shape with a turned-in angle a *concave* shape. It's sort of caved in. Shapes that aren't caved in, that stick out, are called *convex*.

Heather: The turned-in angle is a right angle.

Kevin: But the right angle is outside. The angle on the inside is so big, it's as big as three right angles. There's one from each square, that's two. And then there's two halves of a right angle from the triangles. That makes three.

Good figuring. The caved-in angle is larger than a right angle. Can you see a right angle inside the figure?

Amir: Yeah. There are two on the left side—one on top, one on the bottom.

So we agree about Mei-Ling's shape. How about Jeff's?

Noah: It has five sides, but no parallel sides.

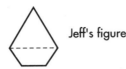
Jeff's figure

Jeff: Yes, it does. This slanty one on the top right and this one slanting the other way on the bottom left.

Does it have exactly one pair of parallel sides?

Amir: No, look. *Two* pairs, here and here. But these angles aren't right angles.

Noah: OK, I agree. Jeff's doesn't follow the rules.

What about my figure?

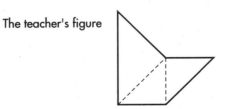
The teacher's figure

Noah: Yours doesn't have *any* parallel pairs! It's got to have one. So yours is no good either.

Well, I thought I followed the rules. Can I get anyone to agree with me?

Manuel: Ah! Look! Here and here *[pointing to the two horizontal sides]*. They're not exactly across from one another, but they do go in the same direction.

[The class comes to a consensus that the teacher's figure does indeed follow the rules.]

So, we found out that parallel sides don't always have to be across from each other. That's not part of the rule. Also, nothing said we couldn't use a turned-in angle. Do you remember what that is called? Sometimes, we assume that there is more to a rule than there actually is.

Which Are Impossible?

These students are discussing the triangles on Student Sheet 9, Can You Make These Triangles?

Cara: This one [tri2] was different.

Tai: Yeah. It has to be a triangle?

Jasmine: There's no such thing!

How do you know? How can you prove it?

Jeff: A triangle is like diagonal, and a right angle is straight.

So does that mean there can't be a triangle with even one right angle?

Antonio: No, there can be one right angle. That's the right triangle. That was one of the ones we had to make and there's a piece... shape L has one, and so does... shape E. But there can't be *two* right angles.

Maricel: You can do it, but it wouldn't be a triangle.

Sofia: Yeah. By the time you make two right angles, you're not going to be able to close it. We tried a whole bunch of ways. *[She refers to her paper and sketches a few.]*

Two right angles next to each other [a], and across from each other [b], and going off in opposite directions [c], but you can't close any without making it not a triangle.

Tai: That's kinda the same as why you can't have a triangle with two parallel sides. Because it won't be a triangle any more if you connect them. You have to connect the top and the bottom, and you can't do that and still have just three sides.

Does anyone have another way to prove that it's impossible for a triangle to have two right angles?

Becky: Three angles have to add up to 180. If you have two 90's, then you reached your limit.

But you still have an angle left and you can't have it be 0.

Almost everyone said that one of the quadrilaterals on the student sheet was also impossible. I'm wondering why.

Marcus: At first we thought the quadrilateral with no pairs of parallel sides [quad5] was impossible until Amir came up with this:

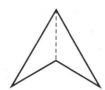

Then we saw it was impossible to build a quadrilateral with four right angles and no equal sides [quad 3].

Jasmine: If it has four right angles, then it has to have sides equal.

Julie: But that doesn't explain why, that's just saying it again.

Katrina: This is hard to explain. If there are 4 corners and they are all right angles, then they all have to be directly across from each other so that they'll connect. If they're not directly across, then you can't draw a straight line to connect them. It has to be diagonal, and if it's diagonal, then it's not 90°. So this can be a square or a rectangle.

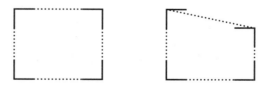

Jeff: I think quad2 is impossible, too. If it has four equal sides, it's a square, so then it has to have right angles.

Toshi: What about this one? *[She holds up shape M, a rhombus.]*

Katrina: Yes, we made one like this *[holding up shape G]*. See, the sides are all the same, but it's not a square.

Using Move and Turn Commands

Materials

- Computers
- Overhead projector and blank transparency, or chart paper
- Students' work on Student Sheets 9 and 10
- Student Sheet 12 (1 per student)
- Student Sheet 13 (1 per student, homework)

What Happens

Students work with *Geo-Logo* move commands (fd and bk) and turn commands (rt and lt). They write procedures that draw a square and a rectangle, and see how they can use the repeat command to write the same procedures in a shorter form. On computer, students use *Geo-Logo*'s move and turn commands to draw an equilateral triangle (and, if time permits, other shapes). In an off-computer assessment, students answer questions about the hierarchical categories of polygons; for example, why a square is also a rhombus. As a follow-up to the computer work, students discuss the difference between turns and angles. Their work focuses on:

- learning *Geo-Logo* fd, rt, and repeat commands and the Label Turns tool
- determining turn sizes for making an equilateral triangle
- distinguishing between turns and angles
- seeing relationships between turns and angles
- defining categories of triangles and quadrilaterals more precisely

---

Activity

Learning the Move and Turn Commands

Note: Students who have done the *Investigations* grade 3 unit *Turtle Paths* or the grade 4 unit *Sunken Ships and Grid Patterns* have had some experience with the commands and procedures reviewed here. If your students have not worked much in *Geo-Logo,* you may need to take more time introducing this activity. Look through pp. 144–146 of the *Geo-Logo* Teacher Tutorial for an overview of the basic commands.

For the *Geo-Logo* activity Polygons with Moves and Turns, which students will be doing for the next two sessions, briefly introduce the *Geo-Logo* commands fd (forward), bk (backward), lt (left turn), and rt (right turn). Students could stand up and follow some of your directions, as if they were the turtle. You might use the computer to demonstrate the *Geo-Logo* commands, particularly if you have a large screen. If you don't have a large screen, illustrate what the turtle does, especially for turns, by drawing the moves on the board or overhead.

Today we're going to look at some *Geo-Logo* commands you may have used before, commands that tell the turtle to move or turn. The move commands are "forward" and "back" *[write* fd *and* bk *on the board]*, and the turn commands are "right" and "left" *[write* rt *and* lt *on the board]*.

Each command is followed by a space, and then a number that tells how far to move for fd and bk, or how much to turn for rt and lt. How should I write a command that tells the turtle to walk 20 steps forward? (fd 20)

Suppose I am drawing a shape, such as a square, that has a right angle in it. What is a command that tells the turtle to make the correct size turn? (rt 90 or lt 90)

Present several other turns:

Suppose the turtle is facing this direction *[draw an arrow pointing up]*, and wants to turn half as much as a turn for a right angle. What is a command for that? (rt 45 or lt 45)

Which direction will it turn after the command rt 45? What is a command that will turn the turtle around to face the opposite direction? (rt 180 or lt 180) What is a command that will turn the turtle all the way around to face in the same direction again? (rt 360 or lt 360)

For an overview of the relationship of angles to turns, see the **Teacher Note**, The Rule of 180° (p. 68).

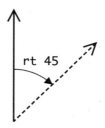

How can we use the commands fd, bk, rt, and lt to direct the turtle to draw a square? Somebody suggest a first command.

Record the command on a piece of chart paper, the board, or an overhead transparency. Continue to ask different students to offer commands until you have an entire procedure for a square. (If students are familiar with the commands to draw a square, you might instead ask a single volunteer for an entire procedure.)

Explain that when we are having the turtle draw shapes on the computer, the turtle should always end up facing the way it started. Thus, even though the square looks complete, the procedure will need to end with a 90° turn.

Learning the Repeat Command

Ask students to think about each command that is given, and to tell what part of the square that command will construct. For example:

```
fd 40
rt 90
fd 40
rt 90
fd 40
rt 90
fd 40
rt 90
```

We just wrote a procedure that will draw a square. Now talk with your partner about how we could write a procedure that would draw a rectangle that is *not* a square.

After a minute or two, take a student's suggestion. Record the rectangle procedure next to the square procedure, and ask students to look for parts of the procedures that repeat.

***Geo-Logo* has a special command we can use if we want to repeat a group of commands. It's called the** repeat **command. This is a procedure that uses it** [write on the board or overhead]:

```
repeat 2 [fd 40 rt 90]
```

How do you think this command works? What do you think it will draw? (two sides of a square, or an upside-down L) **How do you know?**

Students should understand that the repeat 2 instructions mean "carry out the commands that are in the brackets two times." Draw students' attention to the brackets, and explain that *Geo-Logo* will understand this command *only* if brackets are used.

Now work with your partner to rewrite the square procedure and the rectangle procedure we just wrote, this time using the repeat command.

Circulate as pairs work. Share and post the new procedures with the originals.

```
Procedures                    Procedures
to draw a                     to draw a
 square                        rectangle
_____                     _____

  fd  40 ⎫                      fd  20 ⎞
  rt  90 ⎬                      rt  90 ⎟
                                fd  50 ⎬
  fd  40 ⎫                      rt  90 ⎠
  rt  90 ⎬

  fd  40 ⎫                      fd  20 ⎞
  rt  90 ⎬                      rt  90 ⎟
                                fd  50 ⎬
  fd  40 ⎫                      rt  90 ⎠
  rt  90 ⎬

repeat 4 [fd 40 rt 90]        repeat 2 [fd 20 rt 90
                                        fd 50 rt 90]
```

Remember how hard it was to draw tri3 (the triangle with three equal sides and three equal angles) using `jumpto` **and** `setxy`**? You'll be able to do it much more easily with these new commands.**

Classroom Management

Once students understand the use of the repeat command, introduce the on-computer activities for the next two sessions. Students should be getting started on them about halfway through Session 6.

Assign half the class to work on the computer for the rest of Session 6. Groups switch at the beginning of Session 7 and work for the first half of that session.

Groups not on the computer will be working on the Assessment activity, Shapes That Fit Many Categories.

On Computer

Polygons with Moves and Turns

Students open the Polygons with Moves and Turns activity in *Geo-Logo*. They take their work on Student Sheets 9 and 10 to the computer with them.

When you tried to make these shapes on the computer before, using the setxy command, you found that tri3, an equilateral triangle, was difficult to make accurately. Today, you will use the move and turn commands we just talked about, fd (forward), bk (backward), lt (left), rt (right), and repeat, to make an equilateral triangle.

If you have time, see if you can also make quad2 (a rhombus) and tri5 (a triangle with one angle larger than 90°) using these move and turn commands.

Students might start or otherwise mark on their student sheets the specific shapes they are to try. If some students believe that tri2 (two right angles) and quad3 (three right angles with no equal sides) may be possible with the move and turn commands, encourage them to try.

Remember, the computer has tools to help you. You already know about jumpto and the Label Lengths tool. There's also a Grid tool that turns the coordinate grid on and off. (You may want the grid to help you see where you would like to jump to.) And the Label Turns tool tells you the size of the turns in a shape you have made.

 Grid   Label Turns

Refer students to the *Geo-Logo* User Sheet to see these tool icons.

While Students Are Working On-Computer Since the turtle always starts at (0, 0), most shapes students make will be in the first quadrant of the coordinate grid. If students want to make a larger shape, they can use the jumpto command to start the turtle in a different location. Show them how the Grid tool can be clicked on or off as they look for an appropriate location.

You might show students how to use two more of the *Geo-Logo* tools: the Turtle Turner and the Ruler.

 Turtle Turner   Ruler

The first of these tools will help students who are having difficulty making the correct size turn. Clicking on it makes the turtle on the screen into a Turtle Turner, similar to the transparent tool (see p. 184) students will be using in Session 8. With Turtle Turner on, an arrow that begins at the turtle's head will point in the direction the turtle is facing, and rays will appear from the turtle's body in increments of 30° from the arrow. This way students can estimate an angle by skip counting by 30 (for each ray) to figure out approximately how big a turn they want to try.

The Ruler tool will be useful in drawing a figure like tri5, the triangle with one angle larger than 90° and no equal sides. Once students have set the first two sides and found the second turn with the Turtle Turner, they can use the Ruler to determine the length of the final side.

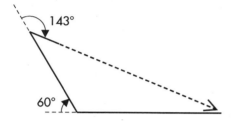

Observing the Students Look for the following as students work:

- What strategy do students use?
- How do they adjust commands appropriately for the feedback the computer gives them?
- Do students check themselves to prove their triangle is equilateral by using the Label Lengths and Label Turns tools?
- Do they know to make a turn larger in order to make the angle smaller? Listen as students talk about turns and angles so you can bring their ideas into the class discussion later.

For students working off the computer, distribute Student Sheet 12, Some Shapes Fit Many Categories. Allow a minute or two for students to read the first question and think about how they might answer it. If you think students need some discussion about how a defined shape can also fit into a broader category, you might say the following:

Here is a similar question that isn't on your sheet: "A square is also a parallelogram. How can that be?" Think about and talk with your neighbors about this. What must be true of a shape for it to be a parallelogram? (four sides, opposite sides equal and parallel) **Does a square have all those attributes?**

Assessment

Shapes That Fit Many Categories

Students may think that to be a parallelogram, a shape cannot have right angles. While it may be true that most shapes they see that are called parallelograms—including the one on the sheet—don't have right angles, all that is necessary for a parallelogram is four sides in two parallel pairs. Parallelograms *with* right angles aren't usually called parallelograms only because they have a more specific name: rectangle, or square.

While Students Are Working Off Computer   Students might do this assessment alone, or they could work in pairs. One possibility is to allow them to discuss the first question together and then ask them to continue on their own.

If students can't think of what to write, ask them to show you what must be true of a rhombus (4 equal sides), and ask if that is always true of a square. Then suggest that they can write down what they have told you.

Some students may have difficulty understanding the questions. Others may think the answers are so obvious that there is nothing to write. When you notice an answer that is essentially correct but incomplete, ask the student to write a bit more or to draw a picture to make the meaning clear.

Most students will be able to answer the first three questions. The fourth and fifth questions are more challenging, as they involve thinking about the way angle size affects side length. Making the equilateral triangle on the computer can help students think about this. Suggest that students who do their computer work after this activity reconsider problem 5 after they have had a chance to make an equilateral triangle.

If some students need a challenge, ask them to think about the converse of the statements on the sheet:

1. Are all rhombuses squares?
2. Are all parallelograms rectangles?
3. Are all isosceles triangles equilateral?
4. Are all scalene triangles obtuse?

Students answer these questions on the back of their paper, explaining their answers.

For some ideas about interpreting students' work on this assessment, see the **Teacher Note**, Assessment: Shapes That Fit Many Categories (p. 66).

❖ **Tip for the Linguistically Diverse Classroom**   Do this assessment orally with students who are not proficient in English. Encourage them to show their thought processes by pointing to specific parts of the drawings on Student Sheet 12 and by drawing their own examples.

How Did You Make the Equilateral Triangle?

Halfway through Session 7, bring the class together again for students to share what they have done on the computer in Polygons with Moves and Turns.

Collect two or three procedures students wrote to make an equilateral triangle in *Geo-Logo.* Ask students to describe the strategies they used.

How are all the procedures different people used to make the triangle the same? How are they different? How do the shapes differ? What must be in a procedure to draw an equilateral triangle? (All turns to make the shape must be 120°, including the final turn to get the turtle facing the same way as when it started, but the side length can vary.)

If enough students made other shapes, you may want to ask similar questions about them.

Are all procedures to make quad2, the rhombus, similar, or are they completely different? (The turns to make the rhombus can vary as long as the sum of two adjacent turns—or of two adjacent angles—is 180°. The side length can vary.)

What turns did you use to make tri5, the obtuse triangle?

Write students' different lists of three turns on the board for comparison.

Students are now finished with their work on Student Sheets 9 and 10. You might want to collect these sheets along with Student Sheet 12 and look them over to see what students have figured out. How are they writing coordinates? Could they all make an equilateral triangle? Students will be making more regular polygons at the beginning of the next investigation.

to make tri 5

fd 30	rt 45
rt 35	fd 70
fd 45	rt 45
rt 166	fd 20
fd 71.7	rt 145
rt 159	fd 85.2
	rt 170

Comparing Turns and Angles

Wrap up Sessions 6 and 7 by taking a few minutes to contrast *turns* and *angles.*

When we first made shapes that followed certain rules, we talked about the *angles* in them. Now we have been using turtle turns. What exactly *is* a turn? What do we mean when we say the turtle turns?

It is important for students to recognize that a turn is a rotation, or a change in direction or heading, without a movement to a different place. For one class's thoughts on the differences between angles and turns, see the **Dialogue Box,** Talking About Angles and Turns (p. 69). The **Teacher Note,** The Rule of 180° (p. 68), discusses confusions that commonly arise.

Draw an isosceles right triangle on the board or on an overhead transparency.

We could use turns to draw this geometric shape on the computer. But we can also talk about the angles in this shape. On this right triangle, who can show us what an angle is? How is it different from a turn?

Invite one or two students up to point out the difference. On the triangle, draw the indicated turn by extending the line of one of the equal sides and showing with a curve and arrow how far the turn would have to go to bring you to the hypotenuse.

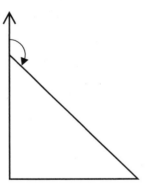

Making turns is like walking along the outline of the shape. Using turns helps us draw a figure on the computer. About how many degrees of turn do I need to make if I want to face along the longest side of this right triangle? (more than a 90° turn, 135°)

Draw another right triangle similar to the first one and indicate the angle inside the turn you just discussed.

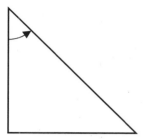

When a shape is already drawn, it is more common to look at and measure the angles on the *inside* of the figure. Like turns, angles are measured in degrees.

About how many degrees do you think this angle is? more than 90°? less than 90°?

Which is larger, the measure of this angle or the measure of the turn that made it? If you want to make a small angle, do you use a big turn or a small turn? (big) **What if you want to make a big angle?** (use a small turn)

One of the things students should realize from their work on these activities is that the turn and the angle together make a straight line, or 180°. Encourage thinking in this direction, but do not push students to see it now. They will have many more opportunities to investigate this idea in the next few sessions.

Sessions 6 and 7 Follow-Up

What Shape Does It Draw?   After Session 7, students work on Student Sheet 13, What Shape Does It Draw? They match each of four procedures with the numbered shapes those procedures would draw in *Geo-Logo*. Then students write the procedure to draw the remaining shape. For the two challenges, students write a procedure to draw a more complex shape, and they predict what more complex shape will be drawn by a given procedure. You may want to suggest that students check the folders of work they have done so far in this unit to make corrections or add to their work as needed.

 **Homework**

Assessment: Shapes That Fit Many Categories

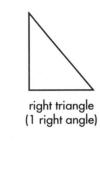

right triangle
(1 right angle)

obtuse triangle
(1 angle larger
than 90°)

scalene triangle
(no equal sides)

isosceles triangle
(2 equal sides)

equilateral triangle
(3 equal sides,
3 equal angles)

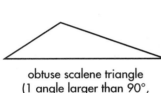

obtuse scalene triangle
(1 angle larger than 90°,
no equal sides)

square
(4 equal sides,
4 right angles)

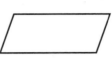

rhombus
(4 equal sides)

parallelogram
(opposite sides
parallel)

rectangle
(4 right angles)

The questions on Student Sheet 12, Some Shapes Fit Many Categories, are designed to help students think about hierarchies of categories of polygons. It takes both an understanding of the definitions of the different categories and logical reasoning to work through the questions. You may find a wide range of student responses.

The strongest answers show both clear knowledge of the characteristics of the shapes and an understanding of how one class of shapes can be included in another. Many answers demonstrate useful knowledge of the shapes and some sense of classification.

Some students write incomplete answers; they may know more than they write down. The weakest answers tend to be either too narrow or too broad. These students might describe the characteristics of a typical example of a shape rather than the shape in general, saying, for example, that a triangle has a point on top; or they might describe a larger category than they are asked for, describing a parallelogram when they are asked about a rectangle, for example.

Following are answers students have given to the questions on the student sheet, with evaluation comments in parentheses. The most complete answers are listed first, without comment.

Question 1. A square is a kind of rhombus. How can this be?

> Because it has four equal sides.
>
> Both have parallel and equal sides.
>
> The sides on the rhombus are tilted. *(incomplete; no mention of equal sides)*
>
> The way it's made is how a rhombus is made. *(too vague; no qualities described)*

Question 2. Name all the shapes above that are parallelograms. How can they be parallelograms and have other names as well?

> All the quadrilaterals. The opposite sides are parallel like a parallelogram.
>
> Square, rhombus, rectangle. Everything is the same except that the angles are different. *(correct shapes, but unclear use of "same")*
>
> Rectangle and rhombus because they have two equal sides. *(too broad; reason can include trapezoids, which are not parallelograms)*

Continued on next page

Rectangle. A parallelogram looks like a rectangle but it is tipped looking. *(incomplete; doesn't mention pairs of parallel lines; doesn't include square and rhombus)*

Rhombus because it's tilted. *(too narrow; considers only one typically pictured parallelogram)*

Question 3. An equilateral triangle is isosceles. How can this be?

They both have at least two equal sides.

They both have equal sides.

Because an isosceles is just a little bit taller than an equal triangle. *(incorrect; too narrow)*

Because maybe they have long sides or the triangle is thinner. *(too vague)*

Question 4. Some obtuse triangles are scalene. Some obtuse triangles are isosceles. Sketch examples of each.

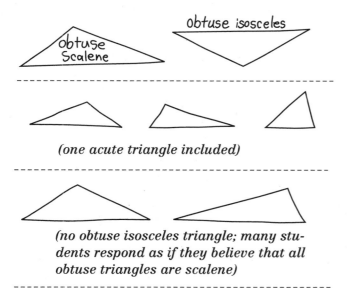

(one acute triangle included)

(no obtuse isosceles triangle; many students respond as if they believe that all obtuse triangles are scalene)

Question 5. Obtuse triangles cannot be equilateral. Explain why this is true.

Because obtuse is one angle bigger than 90°, there can't be 2 bigger than 90°, and an equilateral triangle has all the same angles.

An equilateral triangle is where the vertices are all the same. An obtuse angle cannot fit more than once in a triangle.

Obtuse triangles must have at least one side that is longer.

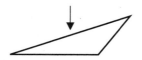

Since all triangles have to have three sides, if you made an equilateral triangle it would not connect.

An obtuse triangle has no equal sides but an equilateral does. *(too narrow; needs to indicate that an obtuse triangle can have no more than two equal sides)*

Understanding angles and angle measures is critical to understanding geometric shapes like triangles and squares. Turtle turning is a powerful and dynamic way to learn about these concepts.

When students are using Turtle Turners, they need to understand the relationship between the angle that the turtle turns and the angle that is formed when the turtle moves forward in its new direction. For example, the following picture shows the turtle's position after starting on the left, then moving forward 100 toward the right.

The next picture shows the position of the turtle and the new direction it faces after it turns 120°.

Below are the results of the turtle moving forward 100 in the new direction.

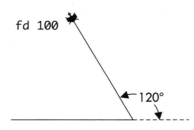

Note that when the turtle moves forward after turning, the angle that it draws is a 60° angle. Even though the turtle has turned through 120°, the lines it draws form a 60° angle.

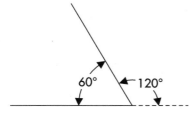

Notice that with 90° turns, the amount the turtle turns and the measure of the drawn angle are the same:

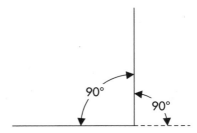

Some students mix up the turn, or exterior angle, and the interior angle. They get confused about the angle through which the turtle is turning. Keep checking to see that your students are visualizing and representing correctly the angle through which the turtle turns (120° in the first set of pictures above), and the angle that is formed (60° in that first set). It is also useful to reinforce that the *greater* the turtle turn or exterior angle, the *smaller* the interior angle.

Another common confusion for students is the relationship of the length of the sides to the measure of the angle. Many students think a larger angle is one with longer sides, or more "area." Thus they believe that of the three angles pictured here, angle B is the largest and angle C is the smallest.

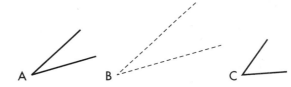

Remind these students that the measure of an angle is the amount of turn from one side to the other. Superimposing pairs of angles shows that angles A and B in fact have the same measure, and angle C has a greater measure than angle B.

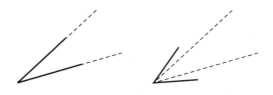

DIALOGUE BOX

Talking About Angles and Turns

These students are watching their teacher demonstrate the computer activity Polygons with Moves and Turns. They are talking about the differences between turns and angles.

Think about angles. What is an angle? *[The teacher types in three commands:]*

```
fd 50
rt 135
fd 50
```

Where is the angle in that?

Noah: The turn is 135°.

If I were the turtle, what would I do when I got to the end of the first `fd 50`?

[Rachel draws an arrow in the direction that the turtle would travel after turning.]

The turtle turns 135°. What do you think the angle is?

Christine: I know, 45.

OK, Christine, show us why you said that.

Christine: I would take 135 away from the half of a circle, which is 180.

If I went up and turned around to face back the way I came, it would be 180°, half a circle. What does Christine's theory say for us?

Matt: I think 135 plus whatever equals 180. Whatever is the angle.

Is there another way to say that?

Rachel: 180 minus 135 is the angle.

Corey: So if there is a turn of 160, the angle is 20. Turn plus angle equals 180.

What is the difference between a turn and an angle?

Jasmine: One's straight and one's curved. You turn around a corner.

Amir: A turn, you have to turn a certain distance, but an angle… just is.

Greg: The turn is what you're actually doing. It's outside. The angle's what's inside.

So Greg's saying the turn's outside and the angle's inside. How else are they different?

Mei-Ling: I think 90° is a turn and 30° or 60° is an angle.

Jasmine: I sort of agree with Mei-Ling. An angle is smaller than 90°.

Why is 90° different from all the others?

Mei-Ling: Because you just turn your body, you don't move in a circle.

Who's not sure? Do you see anything turns and angles might have in common?

Noah: Both can have 90°.

Jasmine: They are the same. 90° turns and 90° angles are the same.

Matt: They are the same except for the name.

Christine: If they have different names, they're different things.

Pretend I'm the turtle. I walk forward 5 steps, then I make a 90° turn. OK, I turned… but did I make an angle?

Greg: No, you'd have to do another forward to have an angle.

So, to have an angle, what do we need?

Amir *[excitedly]*: Two lines!

So, Mei-Ling, do you still think that 90° is a turn, and a different number is an angle?

Mei-Ling: No, you have to have a vertex, or otherwise it's not an angle.

Students are making sense of turns and angles in their own way. The teacher keeps the conversation open to hear different opinions. Perhaps Mei-Ling is not sure yet if angles can be larger than 90°, but there will be more chances for discussion.

Finding Angle Sizes

Materials

- Power Polygons (1 bucket per 6–8 students)
- Overhead projector, blank transparency, and pen
- Student Sheet 14 (1 per student)
- Turtle Turners (1 per student)
- Student Sheet 15 (1 per student, homework)

What Happens

Students use a number of different strategies to find the sizes of angles in the Power Polygons. They discuss their strategies, and they learn to use Turtle Turners as a way of checking their accuracy. For homework, students draw angles by estimating the size, then check each one with a Turtle Turner. Student work focuses on:

- estimating and measuring the size of angles and turns
- using known angles to find measures of other angles

Activity

Finding Angle Sizes

Distribute sets of Power Polygons for students to share (1 bucket per 6–8 students). Place an isosceles right triangle (shape D) on the overhead and trace around it on a transparency.

Look at this isosceles right triangle, shape D in the Power Polygons. Can you figure out the measure of each angle, using nothing but the other Power Polygons in the set?

Students work in small groups; then they share their results and their strategies with the class. Students may give responses like these:

There is a 90° angle because it's a square corner.

We put two shape D pieces together to make a square, so the small angles are 45°.

The two smaller angles are the same size, and together they fit on top of a 90° angle. They're half the 90° angle, so they're 45°.

When the class agrees on the angle sizes, write 90° in the right angle and 45° in each of the smaller angles.

Another way we can prove that the two smaller angles are 45° is to put a number of them together around a point.

Draw a dot on the overhead. Take five of shape D and place them around in a circle with a 45° vertex of each touching the dot.

How many of these angles will fit all the way around?

You might draw one or two lines (as shown in the diagram) to help students see how many more are needed.

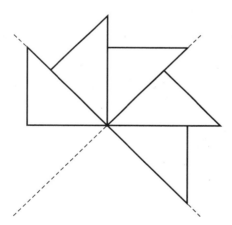

Once you have established that it will take eight triangles to surround the dot, ask students how that could help them figure out how big each angle is, in degrees.

We just made a circle around the dot in the middle. How many degrees are in a circle? (360°) If it takes eight of these triangles to go around the circle, how many degrees must each angle be? How do you know? How many 90° angles fit around the circle?

Don't be concerned if some students remain unconvinced by this method; they will develop their own ways to find angle sizes.

Teacher Checkpoint

Angles in the Power Polygons

Distribute the two pages of Student Sheet 14, Angles in the Power Polygons, to each student. Students work in groups of four.

With your group, figure out the measure of the angles in each of these Power Polygons. You may use other Power Polygons and any information or strategies that you already know. Write the angle measure in each angle in the pictured polygons.

While students are working, circulate and ask groups to show or tell you more than one way that they can figure out a certain angle size. The **Dialogue Box,** Finding Angle Measures of Power Polygons (p. 74), illustrates the many strategies one class used for proving their measures.

Use your observations and the class discussion as a checkpoint:

- Do students understand where the angles are?
- Have all students figured out ways to find angle measures? What are some of these ways?
- How do students build on their knowledge of some angle measures to find other angle measures?
- How do students use the method of grouping angles to form a straight line (180°) or a circle (360°) to find angle measures?

If some students finish early, challenge them to see what other angles they can make by combining angles in the Power Polygons. They can draw these angles on the back of the student sheet. For example, they know each angle of a square is 90°, and the smaller angles in the right triangle are 45°. So if they put them together (shape A next to shape E), how many degrees do they have? (135°)

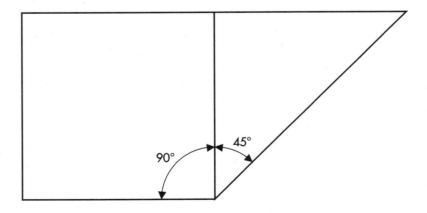

Toward the end of Session 8, call the class together to share their strategies for finding the angle sizes of the Power Polygons included on Student Sheet 14.

Choose for discussion several shapes with angles of different sizes, such as N, L, J, M, K, O, and H.

Estimating Angles   Spend the last few minutes of class introducing the transparent Turtle Turners. Tell students you are going to try to draw a 30° angle, just by looking. On the overhead or the board, use a straightedge to draw an angle that you estimate to be 30°. Then take a transparent Turtle Turner and place it over your angle, with the center of the turtle at the vertex and the arrow along one of the rays of your angle. See how close you came to 30°. Students will be doing something similar for homework.

Discussion: How We Found the Angles

Session 8 Follow-Up

Estimating Angles   Give each student Student Sheet 15, Estimating Angles, and one transparent Turtle Turner to take home. For homework, students do what you demonstrated in class: Draw an angle of a certain measure, using a ruler or other straightedge, by estimating the size. Then they use the Turtle Turner to check their accuracy. They try to draw angles with measures of 30°, 45°, 60°, 90°, and 120°. Students will challenge another person at home to try estimating angles as well. You might remind students to use whole numbers in increments of 5 for the angles they draw.

 **Homework**

Students will be using their Turtle Turners at home again in Investigation 3, so they should find a safe place at home to keep this tool.

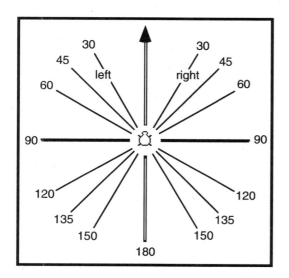

━ D I A L O G U E ☐ B O X ━

Finding Angle Measures of Power Polygons

This class has been working on Student Sheet 14, Angles in the Power Polygons. The teacher assigned one or two shapes to each group; they found the angle measures and now they are presenting their strategies to the class. Students are encouraged to show more than one way to find the measures. Note that the shaded areas in the drawings are parts of shapes that are covered by other shapes

Lindsay: N is an equilateral triangle. One way we found the angles was, we used Leon's rule that there's 180° in a triangle, and we said they're all equal, so we divided 180 by 3 'cause there's three same angles, and you get 60.

Natalie: We also remembered that on the computer we had to turn 120, so the angle is 60. 180 minus 120.

Leon: We did F, and one angle of F is 90 and the other two are equal. To get 180, you do 90 plus 90. So split one of the 90's into 45 plus 45, and you get 90 plus 45 plus 45 equals 180.

Anyone have another way to figure the angles in F?

Shakita: We did the circle thing. *[She places 8 F shapes in a circle around a single point.]* There are eight pieces around and 360° all the way around. So 360 divided by 8 is 45° in each angle.

How did you do J?

Duc: For J, it took 3 of the top angle to make 360°, so it is 120. The other sides had to add up to 60 and they were even, so 60 divided by 2 equals 30.

Matt: To find the angles of J, we used shape L because it fits on half of J. The little angle of L is 30, so the two side angles of J are 30 and that's

60 altogether and there's 180 inside, so 180 minus 60 is 120. The large angle of J has to be 120.

Heather's group, show us something you found out.

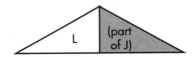

Toshi: For M, we used J to measure. For the other two angles in M, it takes 2 of them to fit exactly in the 120 in J. So divide it by 2 and get 60.

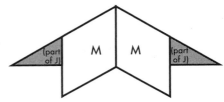

Heather: Just like you put those eight triangles together, we can put these fat angles in M together and make a whole piece with three, so that makes the angle 120, because three times 120 makes 360.

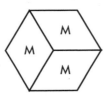

And see, *[moving six of the small angles on M together into a circle]* here there are six. And then 360 divided by 6 makes 60°.

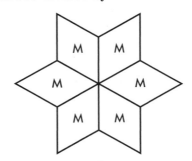

Continued on next page

Continued

Danny: And look *[showing rhombuses M and G],* these are the same shape, aren't they? G is just bigger. The angles are the same for both, the sides are just bigger. See, if you do this *[laying shape M on top of shape G, aligning the 120° angles first and then the 60° angles],* the angles are the same. Also, you could fit four of the small ones inside the bigger one. So the angles are the same, but the sides are different.

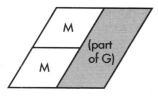

Amy Lynn: For K we started off with this, which makes half a circle, with three *[places three trapezoids around a point, forming a straight edge].* So that makes 180 divided by 3, which is 60.

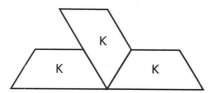

And then, 360 divided into 120's is 3, since you can arrange the big angles into a group of three, too, that make a whole circle.

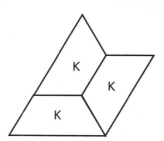

Mei-Ling: Here's how we did O. *[She places six shape O pieces together in a half circle]* Here are six. I'm laying together six of them to make a half circle and here would be six more, so twelve would be a full circle. Now 360 divided by 12 makes 30. So this side *[pointing to the acute angle],* is the angle that is 30.

Now the other small angle is also 30. *[She flips them around to show that both sides fit the same way.]*

But the last two angles of this one are harder. We put two of them together, then we tried to put another one into the space that was left, but it wouldn't fit.

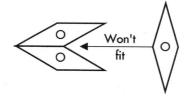

Alani: So we put in the small corners from two more O's, and then they fit. So it took two 30's to finish the whole circle, and 360 minus 60 is 300, and then there's two angles left that are the same. Half of 300 is 150. So those other angles are 150 each.

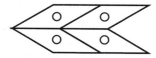

In the process of finding angle measures in the individual shapes, students are finding relationships among the Power Polygons. Danny notices some pairs of shapes that are similar (they have the same angles, and their corresponding sides are in proportion to each other). Other students find the same-size angles in different shapes.

Angles and Turns Together

Materials

- Power Polygons
- Turtle Turners (1 per student)
- Student Sheet 16 (1 per student)
- Computers
- Rulers (1 per 2–3 students)
- Student Sheet 17 (1 per student, homework)

What Happens

Students play a game on the computer that relates the size of an angle to the size of the turn made to form that angle (the supplement of the angle). Off the computer, they estimate and draw some of the angles in the Power Polygon sets, and they use Turtle Turners to measure turns and angles in triangles. Student work focuses on:

- distinguishing between turns and angles
- seeing relationships between turns and angles
- estimating and measuring the size of angles and turns

Activity

On Computer

Angle and Turn Game

Begin by introducing the *Geo-Logo* Angle and Turn game to the whole class. Students will work on this short activity in pairs (it is a two-player game). The game helps students see the relationship between angles and turns.

Demonstrate the game at the computer. Open *Geo-Logo* and choose the Angle and Turn Game. To begin a game, Player 1 enters start and presses the **<return>** key. The players will then see the turtle move to a random location, draw a line segment along its path, and draw a dotted ray to mark its heading.

Player 1 enters a turn command, such as rt 120. The turtle shows this turn and, for this game only, draws a ray to show the new heading. On color monitors, this ray is green for "go."

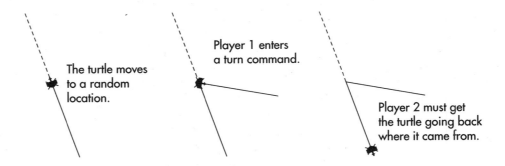

The turtle moves to a random location.

Player 1 enters a turn command.

Player 2 must get the turtle going back where it came from.

The challenge is for Player 2 to figure another command that will turn the turtle to face its starting point (that is, continue turning the turtle through the angle). A dialogue box appears to tell how far off (in degrees) Player 2 was. If the turtle turns through the angle exactly, Player 2 is "off by 0 degrees"—a perfect turn.

Players then switch roles. Whoever was off by fewer degrees wins the round.

This is a brief activity. Once students figure out the relationship between the initial turn (x) and the second turn ($180 - x$), they can always "win." This activity gives students another visual demonstration of the idea that the measure of the turn and the measure of the resulting angle are supplementary; that is, they total 180°.

Activity

Off Computer

Measuring Turns and Angles

Each student needs a copy of Student Sheet 16, Angles and Turns, and a transparent Turtle Turner. Explain that both rows of diagrams on the student sheet show the same three triangles. Using the Turtle Turner, students are to measure the *turns* in the top row and the angles in the bottom row.

Circulate as students work, assisting any who need help. Some may have difficulty figuring out how to position the Turtle Turner to measure an angle as opposed to a turn.

How do you decide where to measure a turn and where to measure an angle? What do you notice about how an angle and turn fit together?

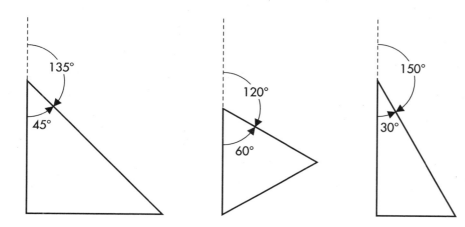

Off Computer

Estimating Angle Size

Students who finish early can work in pairs or small groups to estimate angles. They need one of each Power Polygon shape.

One student picks a shape but does not show it to the others. That student suggests an angle measure that can be found in the chosen shape. For example:

I'm thinking of a 60° angle that is in an equilateral triangle.

The other student(s), using only a ruler and pencil, try to draw the angle. Students talk about how they are planning their drawing. After their angles are drawn, they check them with a Power Polygon or, if they like, a Turtle Turner and correct the drawings as needed.

Students take turns choosing angles in the Power Polygons for the others to draw. For example:

I'm thinking of a 120° angle that is one of the big angles in the trapezoid.

I'm thinking of the 30° angle in the skinny rhombus.

Classroom Management

The main work of this session is the off-computer activities, but students will enjoy and benefit from taking a turn at the Angle and Turn Game on the computer. Each pair should spend a short time playing the game.

You might want to introduce the on-computer activity and the work on Student Sheet 16 to the whole class, and then send as many pairs as you have computers to play the game. The rest of the class can work on the off-computer activities and take turns playing the game as computers become available. Another option is to teach the game to a few students. Then, as each pair finishes, those students teach the next pair how to play.

Demonstrate the Estimating Angle Size activity to groups as they are ready. Save 5 or 10 minutes at the end of class to introduce the homework.

Writing About Angles

Toward the end of Session 9, introduce the homework writing assignment and allow students a few minutes to start it in class.

Many people know about 90° angles, but by now, you are familiar with lots of other angles, too. What are some of the angles you can picture?

Collect students' answers on the board. From their work with Power Polygons, they should feel that they know 30°, 45,° 60°, and 120°.

Write 45° and 60° on the board. Explain that for homework tonight, students are going to write about these two angles. Ask what they might say about them. Take a few suggestions and write them under the appropriate angle. Here are some questions to get them started thinking:

Where do we see a 60° angle? Where do we see a 45° angle?
Who can draw sketches of some shapes that have these angles?
Can you think of any ways that shapes with these angles are used?
How does a 60° angle compare with other angles, such as 90° or 180°?
How does a 45° angle compare with these other angles?
How many of each angle do you need to make a full circle?

Students work on Student Sheet 17, What Do You Know About 45° and 60° Angles?

❖ **Tip for the Linguistically Diverse Classroom** Students who would have difficulty with this amount of writing in English might express their thoughts about 45° and 60° angles primarily through diagrams and mathematical symbols. For example:

$45° \times 8 = 360°$ $60° + \frac{1}{2}$ of $60° = 90°$

Session 9 Follow-Up

What Do You Know About 45° and 60° Angles? Students take home the writing they have begun about 60° and 45° angles on Student Sheet 17, What Do You Know About 45° and 60° Angles?, and continue it for homework.

 **Homework**

Regular Polygons and Similarity

What Happens

Sessions 1 and 2: Regular Polygons Students list the attributes of regular polygons and nonregular polygons, using a few examples as a reference. They find the turns and angles and their sums for regular polygons with from three to ten sides. In the process, they write procedures that draw regular polygons using *Geo-Logo*.

Session 3: Patterns of Angles and Turns Students pool their discoveries about relationships between the number of sides in a regular polygon and the size of its turns and angles. In an assessment activity, they examine three figures made from Power Polygons and tell in writing how they know if each is a regular polygon.

Session 4: Building Similar Shapes Students discuss the mathematical meaning of *similarity* as they look at similar triangles. They pick one shape from the Power Polygons and use several of those pieces to build larger and larger shapes similar to the starting shape. They look for generalizations about what happens to the area of a shape when the lengths of all its sides are doubled and tripled.

Sessions 5 and 6: Similarity Activities On the computer, students write procedures that draw rectangles and houses that are mathematically similar to those provided. Off computer, students investigate similarity with nonregular shapes. They put several Power Polygons together to make an original polygon and build larger shapes that are similar to it. For a final project, they make a poster that shows their similar shapes.

Mathematical Emphasis

- Distinguishing between regular and nonregular polygons
- Exploring the relationship between the number of sides a polygon has and the sums of its turns and angles
- Exploring relationships among angles, line lengths, and areas of similar polygons
- Comparing areas of shapes

What to Plan Ahead of Time

Materials

- Power Polygons: 1 bucket per 6–8 students (All sessions)
- Computers (Sessions 1–2 and 5–6)
- Calculators: available (Sessions 1–2)
- Students' transparent Turtle Turners from Investigation 2 (Sessions 1–2)
- Rulers: 1 per student (Sessions 1–3)
- Overhead projector (Sessions 1–2 and 4)
- Large paper for posters: 1 sheet per student, plus extras (Sessions 5–6)
- Stick-on notes: 1 pad (Sessions 5–6)
- Colored pencils, markers, or crayons (Sessions 5–6)

Other Preparation

- Duplicate student sheets and teaching resources, located at the end of this unit, in the following quantities. If you have Student Activity Booklets, copy only the items marked with an asterisk.

For Sessions 1–2

Polygons: Regular and Not Regular (p. 194): 1 transparency*

Student Sheet 18, Total Turns and Angles (p. 185): 1 per student, and transparency of first page only*

For Session 3

Student Sheet 19, Which Are Regular Polygons? (p. 187): 1 per student

Student Sheet 20, Polygons That Are Not Regular (p. 188): 1 per student (homework)

For Session 4

Student Sheet 21, Building Similar Shapes (p. 189): 1 per student, and 1 transparency*

Student Sheet 22, Length of Sides Versus Area (p. 190): 1 per student (homework)

For Sessions 5–6

Student Sheet 23, Similar Rectangles (p. 191): 1 per student

Student Sheet 24, Similar Houses (p. 192): 1 per student

Student Sheet 25, Drawing More Similar Rectangles (p. 193): 1 per student (homework)

One-centimeter graph paper (p. 195): 2–3 per student (homework)

Computer Preparation

- Be sure a copy of the *Geo-Logo* User Sheet (p. 168) remains posted by each computer.
- Work through the following sections of the *Geo-Logo* Teacher Tutorial:

Guess and Test Polygons   135

Drawing Regular Polygons   136

    How to Draw Regular Polygons; How to Use the Draw Commands and Change Shape Tools

Similar Rectangles   139

Similar Houses   141

    How to Make Similar Houses; How to Use the Scale Tool for Help; How to Use the Scale Tool to Check Your Procedure

Continued on next page

■ Plan how to manage the computer activities, depending on computer availability.

With five to eight computers: Follow the investigation structure as written. Half the students work at the computer in pairs or threes while the other half work off computer. Groups then switch.

With a computer laboratory: Begin Session 1 with the whole-class discussions. Students then complete the computer work and finish the session with the off-computer activities.

Similarly, for Sessions 5 and 6, students work on the computer first and move to the off-computer activities as they finish.

With fewer than five computers: Introduce the on-computer activities in Session 1 and immediately assign some students to begin cycling through them. Make and post a schedule, assigning about 20 minutes of computer time for each pair of students throughout the day. Students can continue the computer work from Sessions 1 and 2 while the class moves on to Sessions 3 and 4.

Similarly, immediately after you have introduced the on-computer activity in Session 5, assign some students to begin cycling through it. Students may have to complete their computer work for this unit as you begin a new unit, or while the rest of the class is doing a Ten-Minute Math activity or other short session.

Regular Polygons

What Happens

Students list the attributes of regular polygons and nonregular polygons, using a few examples as a reference. They find the turns and angles and their sums for regular polygons with from three to ten sides. In the process, they write procedures that draw regular polygons using *Geo-Logo*. Their work focuses on:

- distinguishing regular polygons from nonregular polygons
- finding the sizes of turns and angles for regular polygons
- writing *Geo-Logo* procedures for making regular polygons
- finding patterns for turns, angles, and the sums of turns and angles for regular polygons

Materials

- Transparency of Polygons: Regular and Not Regular
- Students' Turtle Turners
- Rulers (1 per student)
- Computers
- Student Sheet 18 (1 per student, homework)
- Transparency of Student Sheet 18, page 1
- Overhead projector
- Power Polygons
- Calculators

Activity

Display the transparency, Polygons: Regular and Not Regular. Students should have their Turtle Turners and rulers to use as they wish.

What is it about a polygon that makes it regular? What makes a polygon *not* regular?

I want you to take a few minutes to write answers to these questions, on notebook paper, in your own words. Use illustrations if you want to.

After everyone has had a chance to write something, ask for volunteers to read their answers and to show what they mean by pointing to polygons on the transparency.

Students are likely to say that a regular polygon is equilateral (all sides are the same length). It is important for them to recognize that for a figure to be a regular polygon, all its *angles* must be equal as well as all its sides. A star shape may have sides that are all the same length, but its angles are alternately acute and obtuse, so it is *not* a regular polygon.

Defining Regular Polygons

On Computer

Guess and Test Polygons

This is one of two computer activities to be done during Sessions 1 and 2; it is a brief activity that offers reinforcement in the concept of regular polygons.

If you have a projection device or a large-screen monitor for the computer, do the Guess and Test Polygons activity with the whole class at once. If you do not have a projection device, you might divide the class into groups for as many computers as you have and work through the activity together. Or, individuals or pairs might do the activity during free time during another part of the day. If you feel some students do not need this reinforcement, they need not spend much time on the activity.

Open *Geo-Logo* and choose the Guess and Test Polygons activity. Type poly1 in the Command Center, press **<return>**, and a polygon will appear.

Your job is to decide whether or not the polygon is regular and why. If you disagree, take time for discussion and see if you can come to an agreement. If you agree, how could you prove that a polygon is regular or not regular?

If students do not suggest using the Label Lengths and Label Turns tools, remind them where these guides are on the tool bar, and suggest using them to check their guesses and see if the side lengths and turn measures are the same.

Students continue by clicking on the Erase All tool, then typing poly2 (and later poly3, poly4, and so on up to poly10) in the Command Center and hitting **<return>** to see the next polygon. Or, they can use the cursor and keyboard to change the number after poly to call up the different polygons.

Students may want to return to this activity later in the session when they are trying to write procedures that draw regular polygons. Encourage them to use this as a resource, and suggest it to students who are having trouble.

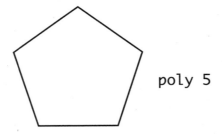

poly 5

Finding Total Turns and Angles

As background for the activity students will be doing both on and off the computer, review the procedure they used to draw an equilateral triangle in Investigation 2.

Do you remember trying to draw an equilateral triangle on the computer? First you tried with setxy commands, and it was very difficult. Later you used move and turn commands, like forward (fd) and right (rt), and you found it was a lot easier. Talk to your neighbor—see if you can remember a procedure you wrote to draw an equilateral triangle.

On the board, record one or more procedures students suggest for drawing an equilateral triangle.

Distribute the Power Polygons among groups and give each student the two pages of Student Sheet 18, Total Turns and Angles. On the overhead, display the transparency of the chart on this sheet. Draw attention to the first row: equilateral triangle.

Give students a few minutes to fill in this row of their chart with a partner. They can ignore the column for *Length of sides* at this time; later they will use it to record the lengths of the sides of the polygons they are making on the computer.

Polygon	Number of sides, turns, angles	Length of sides (turtle steps)	Size of each turn (degrees)	Sum of turns	Size of each angle (degrees)	Sum of angles
equilateral triangle	3		120°	360°	60°	180°

Take a few minutes to discuss this information together. If necessary, review the difference between turns and angles. Check to see that students understand how to fill in the sums of turns and sums of angles.

How many turns need to be added together in a triangle? How many in a hexagon?

❖ **Tip for the Linguistically Diverse Classroom** Point out that each word in the first column of the chart is supported by a number in the second column to remind students of the number of sides in each polygon listed.

Explain that page 2 of Student Sheet 18 pictures each of the polygons on the chart. This is a reference, to show for example what a regular octagon looks like; it can help them when they need to draw one on the computer.

Students will be gathering information to complete Student Sheet 18 as they work on and off the computer in Sessions 1 and 2. Take a few minutes to talk about the on- and off-computer activities and how students will organize their time during these sessions (see Classroom Management, p. 87).

Activity

On Computer

Drawing Regular Polygons

On the computer, students work again in the Polygons with Moves and Turns activity. Earlier they have written procedures to draw an equilateral triangle and a square; now they write procedures to draw regular polygons with five (pentagon), six (hexagon), and more sides. Polygons with seven and nine sides are a challenge, but most pairs will be able to find a close approximation for the turn and angle sizes and exact answers for the totals.

As students work at the computer, they have Student Sheet 18 with them to record information about the polygons they are making. Remind them of the tools available to help them: Power Polygons, their Turtle Turners, the reference pictures on Student Sheet 18, the *Geo-Logo* User Sheet, *Geo-Logo*'s tool bar, and calculators.

Students may want to begin their procedures with a `jumpto` command so they start their drawing somewhere other than the center of the screen. Remind them how to turn the Grid tool on and off, and encourage use of the Label Length and Label Turns tools.

Students who are having difficulty with a particular polygon might try using the Draw Commands tool to approximate the turns by drawing them freehand, perhaps looking at a picture of the polygon for guidance.

 Draw Commands

 Change Shape

Whether they enter their own commands or use the Draw Commands tool, students often get a shape that is roughly, but not exactly, a regular polygon. At that point, they can try the Change Shape tool. This tool lets them use the mouse to drag and move sides, changing the shape of a figure in the Drawing window. The commands also change automatically as the shape is changed. Refer to the *Geo-Logo* Tutorial, p. 137, for further discussion of these two tools.

The **Dialogue Box,** Drawing Regular Polygons (p. 89), illustrates how students in one class hit upon two key concepts as they worked on this activity.

Students fill in the chart on Student Sheet 18.

Those who do the off-computer work first may use Power Polygons, their Turtle Turners, rulers, and the reference drawings on page 2 to help them identify turns and angles.

If they have worked on the computers first, Student Sheet 18 will be partially complete. They use the other tools available to check the degree measurements for the polygons they drew on the computer. They also find turns and angles for any polygons they didn't draw on the computer.

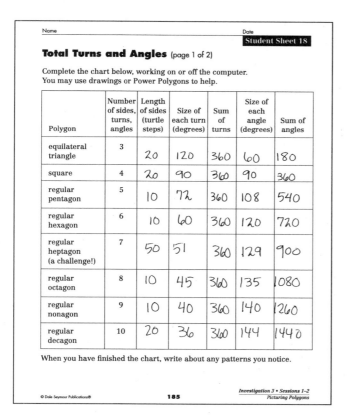

Name _____ Date _____

Student Sheet 18

Total Turns and Angles (page 1 of 2)

Complete the chart below, working on or off the computer.
You may use drawings or Power Polygons to help.

Polygon	Number of sides, turns, angles	Length of sides (turtle steps)	Size of each turn (degrees)	Sum of turns	Size of each angle (degrees)	Sum of angles
equilateral triangle	3	20	120	360	60	180
square	4	20	90	360	90	360
regular pentagon	5	10	72	360	108	540
regular hexagon	6	10	60	360	120	720
regular heptagon (a challenge!)	7	50	51	360	129	900
regular octagon	8	10	45	360	135	1080
regular nonagon	9	10	40	360	140	1260
regular decagon	10	20	36	360	144	1440

When you have finished the chart, write about any patterns you notice.

© Dale Seymour Publications®   **185**   *Investigation 3 • Sessions 1–2*
Picturing Polygons

Classroom Management

There are a number of options for organizing the work on Student Sheet 18, on and off the computer. Students will have the rest of Session 1 and all of Session 2 to work on this task.

Everyone might work in pairs, with half the class starting on the computers and half the class off. Groups switch periodically so that everyone has equal access to the computers.

Another option is for students to work in groups of four, again alternating for equal time on and off the computer; the entire group could be responsible, as a team, for the finished chart on Student Sheet 18. These groups might have access to one computer that is "theirs" for these sessions, or they might rotate from an on- to an off-computer activity after filling in a certain number of rows, which their partners check before working on more rows.

You might let students indicate their preference for starting work on or off the computer. Those who start on the computer may use more trial and error as they try to draw regular polygons and discover the sizes of the turns and angles required. Students who work off-computer first determine turns and angles by another method, then use the computer as a check to see that the numbers they found are correct.

Sessions 1 and 2 Follow-Up

 **Homework**

Total Turns and Angles   After Session 2, students continue to work on Student Sheet 18, Total Turns and Angles, at home. They use the Turtle Turners they took home during Investigation 2 to continue to find angles and turns. They also look for patterns that will help them find the values for a seven-sided polygon if they have not already done this.

Those who have finished the chart might try to extend it, filling in the values for polygons with many more sides, perhaps a 15-sided and a 20-sided polygon. They explain their work on the back of the student sheet.

 **Extension**

Writing Procedures with Angles   There is a feature of *Geo-Logo* that enables students to draw shapes using the interior angle measure rather than turns. Suggest that students try using the commands rta (right turn angle, which makes an angle to the right) and lta (left turn angle, which makes an angle to the left) to draw shapes by specifying the angles.

Start with a forward (fd) move. Then when you enter an rta __ command, the turtle first faces the opposite direction and then turns right the number of degrees specified. With another forward command, the turtle creates the angle.

To make an equilateral triangle, then, we could enter:

```
repeat 3 [fd 30 rta 60]
```

Encourage students to use these new commands to make regular polygons by specifying the angle instead of the turn.

Drawing Regular Polygons

Pairs are working at the computer to write *Geo-Logo* procedures that draw regular polygons. The teacher reminds one pair about the repeat command and then suggests trying some turn sizes that are a bit too small and some that are too large, to see what happens.

If you were doing the pentagon, how many times would you need to repeat what's in the brackets?

Greg: Five.

What if you make a triangle?

[Sofia types in repeat 3 [fd 40 rt 120]*.]*

Sofia: It worked!

What if you change the 120 to 116? *[Sofia does this.]* **What happens? How do we know there's a problem?**

Greg: It doesn't reach.

What does it need? What needs to happen to the size of the turn?

Sofia: It needs to be larger.

OK, try 123.

[Sofia enters the new command.]

Greg: Too big. The angle crosses itself so the turn's too big.

And if it does not meet? If the shape is open?

Sofia: Then the angles are too large, and the turns that made them are too small.

Why do large angles go with small turns?

Sofia: Because together they make 180°.

Sofia sums up a key relationship between turns and angles: The larger one is, the smaller the other, because together they have a sum of 180°.

The next pair the teacher observes are just starting a pentagon.

Rachel: We have to figure how much the turns have to be. The square is 90.

Desiree: The pentagon looks almost 90... maybe 85. *[She tries it].*

Rachel: That's too little.

Desiree: No, it's too much. Because if it were 90, it would make a square *[tracing the figure on the screen with her finger]*. The turn has to be less.

How many degrees did the turtle turn in all to make the square and the equilateral triangle?

Rachel: Wait. Like we did for the triangle. Divide 360 by 5.

[The result of 72° works, and they immediately go on to the hexagon.]

Desiree: 360 divided by 6.

Rachel: 60. Oh, so it goes down every time!

What did you figure out?

Desiree: You have to divide the number of turns by 360, I mean into 360. Because there's going to be six angles, and you need to turn 360 degrees, and that will tell you how much to turn.

Rachel: Not always six angles.

Desiree: Well, I mean how many angles are in the shape.

Why 360?

Rachel: Because that's how many degrees it takes to go around a circle. *[As Rachel says this, Desiree moves her finger in a complete circle on the screen.]*

Desiree: Like the turtle always ends up facing the same direction. So divide how many turns into how much it has to turn to stand up again.

These pairs are grappling with two ideas that together allow us to figure the angle and the turn size in any regular polygon. We can divide 360° by the number of turns to find the share for each turn; then, to find the angle size, we can subtract the turn from 180°.

Patterns of Angles and Turns

Materials

- Completed work on Student Sheet 18
- Student Sheet 19 (1 per student)
- Rulers (1 per student)
- Power Polygons
- Student Sheet 20 (1 per student, homework)

What Happens

Students pool their discoveries about relationships between the number of sides in a regular polygon and the size of its turns and angles. In an assessment activity, they examine three figures made from Power Polygons and tell in writing how they know if each is a regular polygon. Their work focuses on:

- finding patterns for turns, angles, and the sums of turns and angles for regular polygons
- distinguishing regular polygons from other (nonregular) polygons

---

Activity

Discussion: Patterns for Regular Polygons

Students gather in small groups to discuss their findings on Student Sheet 18, Total Turns and Angles. They share their strategies for figuring out the sizes of turns and angles in each of the regular polygons.

Then bring the class together for a whole-group discussion. Direct their attention to the columns for size and sum of turns and size and sum of angles. Did they find any patterns?

Encourage all groups to share. Some groups may have figured out that the total of the turns in each polygon is 360° and used this in a formulaic way to get the size of each turn (360° divided by the number of angles). Some may have worked with the knowledge that the angle and the turn together form a straight line (totaling 180°). Some students may not have used 180 or 360 at all, but instead examined the drawings and shape pieces and used the known measures of angles in the shapes to measure turns.

Were you able to predict the size of the turn and the angle for a 7-sided polygon? How did you figure it out? Can you predict the sizes of the turn and angle for a 15-sided polygon? for a 20-sided polygon? How would you do this?

What is special about 360°? Why is 360 an important number?

Ask students what happens as the number of sides increases in a regular polygon. They should notice that as the angles get larger and the sides get shorter, the figures look increasingly like circles.

Spend the last half of Session 3 on this assessment activity. Distribute Student Sheet 19, Which Are Regular Polygons? Students decide which of the figures are regular polygons and explain in writing how they know. Make Power Polygons available for students who want to build the figures and compare parts of the shapes directly.

Ask students to explain their thinking fully so you will know what knowledge they used to make their decisions. Encourage them to use the diagrams on the student sheet to help make their arguments clear.

❖ **Tip for the Linguistically Diverse Classroom** Do this assessment orally with students who have limited English proficiency.

As you observe students working, and later as you review their papers, consider these questions:

■ Can students determine that shape 1 is a regular polygon and shapes 2 and 3 are not? (Shape 2, the pentagon, has neither equal angles nor equal sides; Shape 3, the octagon, has equal angles but unequal sides.)

■ What kinds of proof do students use? Do they use only appearance statements, such as "It looks like a regular octagon," "It is made up of the same shapes," or "The opposite sides are the same"?

■ Do students recognize that the lengths of all the sides have to be equal in a regular polygon?

■ To compare lengths of sides, do students use tools (such as rulers) or knowledge about the shapes that make up the polygons to prove their statements? For example, for shape 2, one student wrote:

> I compared the top shapes to the bottom shapes, and the edges of the top ones were longer than the bottom ones.

For shape 3:

> The slanted lines are longer than the straight ones. I know because I took the triangle and put it next to the square.

■ Do students recognize that the measures of all the angles must also be the same in a regular polygon?

■ To compare measures of angles, do students use tools (such as Turtle Turners) or knowledge about the shapes that make up the polygon to prove their arguments? For example, one student used her knowledge of the angle size in the isosceles triangle that makes up shape 1 to figure out that "the top corner is 120°, and there are four 30° angles in the others; and $4 \times 30 = 120$. So they are all the same."

■ Do students attend to *either* side length or angle measure, or do they consider both simultaneously? For example, one student said shape 3 "is not [a regular polygon] because even though all the angles are equal, the sides are different lengths so it can't be a regular polygon." Another student wasn't sure:

> I don't know because the sides aren't the same, but the turns are the same and the angles are the same.

Some students provide specific information:

> [Shape 1] is, because the angles are 120° and all the sides are 2½ centimeters.

> [Shape 3] is not, because the angle for all of them is 135°, but the length is 2 cm for the top and the bottom and the two sides, and the corners are 2½ cm.

Students who finish early consider the following question: How is a circle different from a polygon? They write an explanation of what a circle is and how it differs from a polygon with 1000 sides.

Session 3 Follow-Up

Homework

Polygons That Are Not Regular   Students investigate whether the patterns they found for sums of the angles on Student Sheet 18 also hold for polygons that are not regular. You might want to review the pattern that was discovered for regular polygons.

Distribute Student Sheet 20, Polygons That Are Not Regular. Divide the class into four groups and assign two polygons to the students in each group; for example:

> Group A: triangles and decagons
> Group B: quadrilaterals and heptagons
> Group C: pentagons and octagons
> Group D: hexagons and nonagons

Each student should draw at least two examples, with different-sized angles, of the two nonregular polygons he or she is investigating. Students measure the angles of each nonregular polygon using the Turtle Turners they took home during Investigation 2, and add them to find the sum. Students will share their results in Session 4.

Building Similar Shapes

What Happens

Students discuss the mathematical meaning of *similarity* as they look at similar triangles. They pick one shape from the Power Polygons and use several of those pieces to build larger and larger shapes similar to the starting shape. They look for generalizations about what happens to the area of a shape when the lengths of all its sides are doubled and tripled. Their work focuses on:

- recognizing and exploring similar shapes
- creating geometric patterns that grow in regular ways
- exploring connections between geometric and numerical patterns
- exploring the relationships between lengths of sides and areas of similar shapes

Materials

- Overhead projector
- Power Polygons
- Student Sheet 21 (1 per student)
- Transparency of Student Sheet 21
- Student Sheet 22 (1 per student, homework)

Activity

In small groups, students share the results of their investigation of the sums of angles in nonregular polygons. Then they share their results with the whole class. List polygons by number of sides, and write all the different angle sums students found.

What different methods did students use to measure angles? Did anyone measure an angle greater than 180°?

Pooling Homework Results

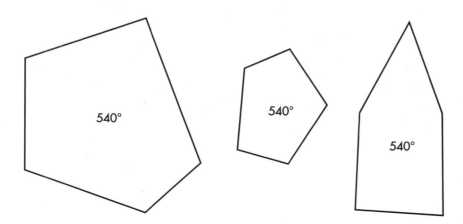

Exploring Similarity

Display on the overhead three similar polygons of different sizes—such as the three isosceles right triangles (D, E, and F) from the set of Power Polygons.

What is the same about these three shapes? What is different?

Students may notice that all the shapes are triangles with one 90° angle and two 45° angles, that all have two equal sides, and all are the same shape. Besides their color difference, students will notice a general difference in size.

You said their angles are all the same size, but the triangles are *different* sizes. What is it about one triangle that makes it bigger than the others? If you were writing a procedure to make the computer draw these triangles, what would be different about them? (the numbers in the `fd` commands)

Introduce the term *similar* as a word to describe these triangles. Explain that shapes can be different sizes but still be called similar. The **Teacher Note,** Similar Shapes (p. 99), further discusses similarity and difficulties students might have with this concept.

All regular polygons with the same number of sides are similar. For example, all squares are similar. But nonregular polygons are not necessarily similar.

Place a rectangle C from the Power Polygons on the overhead. Have several more C pieces available.

Let's say we want to make larger and larger rectangles that are similar to this one, using only rectangle pieces this size. How many should we put together to build the next larger similar rectangle?

Ask a student to come up and build it on the overhead, using only C rectangle pieces. If a student builds a similar rectangle with more than four pieces, ask if that's the next size similar rectangle. If the student builds a rectangle using two or six shapes, ask if it is really similar to the first.

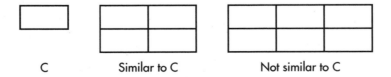

C   Similar to C   Not similar to C

After the rectangle with four pieces is built, build a 2×3 rectangle next to the first two examples. Ask students if they feel that the third rectangle is similar to both the first and second rectangles.

Many may see the third rectangle as similar. Ask students to explain their reasoning, calling on students who think it is similar as well as those who think it is *not*. Continue the discussion until you are sure students understand the concept of similar shapes and realize that the third rectangle is not similar to the first two because it is too wide for its height.

The width of this last rectangle is three times the original; the height is only two times the original. To make it similar, we would need nine pieces altogether, another row of three. If the width is tripled, the height must be tripled too, to end up with a similar figure.

You might demonstrate how to check for similarity by standing directly over a shape that is built, closing one eye, and holding the original shape above it, then moving the original shape up or down to see if you can align all of its sides with the constructed shape. You should be able to line up shape C with the 2×2 rectangle, but not with the 2×3 rectangle. The latter should be too wide. Students may be able to do this at their seats by using one C piece to eyeball the figures projected from the overhead.

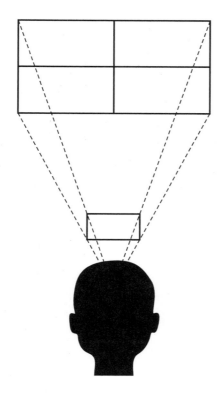

Building Similar Shapes

Hand out Student Sheet 21, Building Similar Shapes, and display the transparency of it on the overhead. Explain that students will be using the Power Polygons to build larger and larger similar shapes, just as they did with rectangle C in the previous activity. Use the rectangle C example to show how to record on the chart, asking students for the number of rectangles needed for the second rectangle you made (4), and for the third rectangle (9).

For each shape on Student Sheet 21, students consider that the smallest similar figure takes one piece—the shape itself. Each time they build the next larger shape, they record the number of pieces it took. Students working together can pool their information for most of their responses, but each student should independently build the similar trapezoids and hexagons.

Warn students that the hexagons are a challenge and will require a different strategy than the rest of the figures. (Students will need to break some of the hexagons into smaller shapes.)

Students need not work on the shapes in the order they are listed on the sheet. In fact, sharing of Power Polygons will be easier if each person in a group starts working on a different shape.

Also, students should know that they can use different pieces if it makes building faster and easier. For example, since the blue rhombus (M) is the same size as two green triangles (N), students may use some blue rhombuses when they are building larger triangles. However, when they count the number of pieces they used, they will need to think in terms of the shape piece they started with. In other words, how many of the *original* shape pieces are needed to cover the new similar shape?

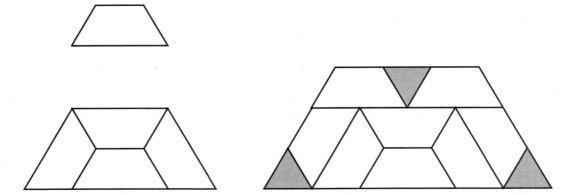

Generalizing About Growing Shapes

Toward the end of Session 4, call the class together to compare their results on Student Sheet 21.

What happened when you used the Power Polygons to build larger and larger shapes? What patterns did you see? Why do you think they occur?

Students may observe that "the numbers were all the same," or "they were square numbers." If any do not know the term *square numbers,* show with square tiles that it's the sequence of the numbers of squares it takes to build larger and larger squares. Students may want to make conjectures about why enlarging shapes other than squares also produces square numbers.

Pay particular attention to the shapes students found more difficult to enlarge. Ask for volunteers to show the class how they put shapes together to enlarge the trapezoid (K) and the hexagon (H). If no one was able to do them, suggest that students consider "breaking" the shapes into smaller pieces: the trapezoid into three triangles (N), and the hexagon into two trapezoids (K) or three rhombuses (M).

Explain what happened when you tried to make a similar hexagon. Did your first attempt work? What shapes did you need to use?

Talk with students about how they figured out how many shape pieces it would take to build the tenth enlargement. Did they have to build it, or could they predict the answer by using the patterns? Can they predict for other enlargements, such as the twentieth?

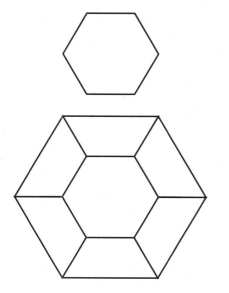

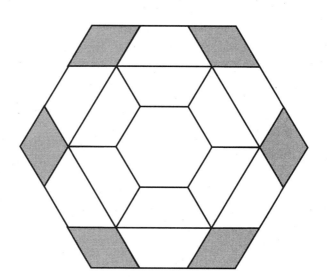

Start a chart on the board that recaps the findings on Student Sheet 21:

When the sides are this many times as long...	The area is this many times as large...
2	4
3	9
4	
5	
6	

Extend the chart as a class, according to students' suggestions.

What patterns are you using to figure out how many of the originals fit? Why does the pattern work this way?

Session 4 Follow-Up

🏠 **Homework**

Length of Sides Versus Area   Distribute Student Sheet 22, Length of Sides Versus Area, for homework.

❖ **Tip for the Linguistically Diverse Classroom** Read the assignment aloud, encouraging students to add simple sketches and mathematical symbols to help them recall the meaning of each question. Students may respond in their primary language or rely on sketches with number labels to answer the questions.

Students' work on this sheet serves as a good checkpoint of their understanding of similarity and the relationship of area to perimeter in similar figures.

Similar Shapes

In general usage, the word *similar* means having characteristics in common. In geometry, the meaning is more precise: Two figures are similar if they have exactly the same shape—if their angles are equal and the sides of one figure are in proportion to the sides of the other. Scale drawings produce similar shapes.

The first three rectangles below are similar; their sides are in proportion. If you double the lengths of all the sides of the 2 × 3 rectangle, you get a 4 × 6 rectangle. If you triple the lengths of the sides, you get a 6 × 9 rectangle. As long as you multiply all the sides of a rectangle by the same factor, you will get a similar rectangle.

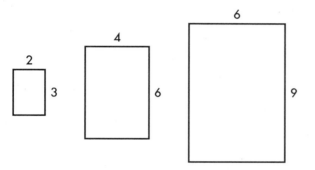

The next three rectangles are not similar. Their sides are not in proportion.

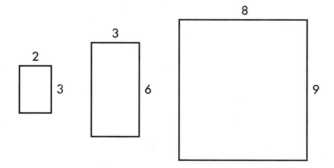

Students often erroneously believe that having the same shape means "described by the same shape name"—*rectangle,* for example. So a "skinny" rectangle and a "fat" rectangle may mistakenly be thought of as having the same shape just because they're both rectangles.

In geometry, figures that are considered to have the same shape must be in proportion, like figures that are duplicated, reduced, or enlarged by a copy machine. So "having the same shape" and "being similar" are equivalent, but "having the same shape" and "being described by the same shape name" are not.

According to these rules, any regular polygon is similar to any other regular polygon that has the same number of sides. For example, any regular, or equilateral, triangle has three 60° angles and three equal sides. If you take an equilateral triangle and enlarge it—or shrink it—to any other size equilateral triangle, it will still have three 60° angles and equal sides. It will be similar to the original one.

If you enlarge or shrink a regular pentagon (five equal sides and five equal angles of 108°) to any other size regular pentagon, the new figure will be similar to the original one. The shape of the Pentagon building in Washington, D.C., is similar to the pentagon on Student Sheet 19.

The activities in this unit build ideas about similarity without requiring formal use of proportion. We use other terms, such as *same shape* and *scale* to discuss the idea of mathematical similarity. When you introduce the term *similar,* alert students to the distinction between its mathematical meaning and its general meaning. If students need additional help understanding the idea of similar figures, you might show examples of similar rectangles and compare them to rectangles that are not similar.

Similarity Activities

Materials

- Student Sheet 23 (1 per student)
- Student Sheet 24 (1 per student)
- Computers
- Power Polygons
- Large paper for posters (1 sheet per student, plus extras)
- Colored markers, pencils, or crayons
- Stick-on notes (1 pad)
- One-centimeter graph paper (2–3 per student, homework)
- Student Sheet 25 (1 per student, homework)

What Happens

On the computer, students write procedures that draw rectangles and houses that are mathematically similar to those provided. Off computer, students investigate similarity with nonregular shapes. They put several Power Polygons together to make an original polygon and build larger shapes that are similar to it. For a final project, they make a poster that shows their similar shapes. Student work focuses on:

- predicting the lengths of the sides of larger and smaller similar shapes
- writing a *Geo-Logo* procedure that draws a figure similar to a given figure
- exploring the relationship between the side lengths and the areas of similar shapes
- exploring connections between geometric and numerical patterns

Activity

On Computer

Similar Rectangles

Distribute Student Sheet 23, Similar Rectangles, and Student Sheet 24, Similar Houses, to each student. Explain that these are recording sheets they will use at the computer for two activities they will be doing.

Demonstrate the Similar Rectangles activity to the whole class, either using a projection device or gathering students around a single computer. Increasing the font size (**All Large** in the **Font** menu) will make your commands more visible for the demonstration.

Students will be using one new tool in this activity:

Scale

Open *Geo-Logo* and click on Similar Rectangles. When you click **[OK]** or press **<return>**, the activity opens with a rectangle procedure in the Command Center and the corresponding rectangle is created in the Drawing window. This is the "original" rectangle referred to on Student Sheet 23. Click on the Scale tool and then, with the hand, click on the upper-right corner of the rectangle. Drag the corner of the rectangle.

What do you notice? How is this rectangle changing?

Students should observe that the figure keeps the same shape exactly and changes only its size, becoming smaller or larger.

Now watch the commands in the Command Center as I drag the rectangle. What do you notice? What is changing?

The turn commands are *not* changing, while the forward commands are changing, all of them getting larger (or all getting smaller). Students may not recognize that the numbers also change in proportion; they will come to appreciate this concept gradually.

Use the Scale tool again to change the rectangle back to its original size (the first command being fd 10). Ask students to predict and record, on Student Sheet 23, the procedure to draw a similar rectangle if the first command is fd 20.

After a few minutes, ask students to explain the reasoning they used. You might hear ideas such as "20 is double 10, so I doubled 15 to get 30," or "15 is 10 plus half of 10, so I used 20 plus half of 20, and that's 30." Be sure they have left the four turn commands unchanged.

If students predict sides of 20 and 25 because they add 10 to each side, ask them to watch carefully as you use the Scale tool again. Stretch the figure so the vertical sides measure 20.

What are the other measures? How do they compare to your predictions? How does the 20 relate to the previous length of 10? What is the length of the other two sides? How does this relate to the previous length of those sides?

If the students' predicted commands were accurate, they put a check mark by the *Predict* heading. If there were any errors in their predictions, students circle those and write the actual command in the *Corrected* column.

When you go off to work at the computer, your first task will be to predict the move and turn commands to make the other three similar rectangles on this sheet. The first fd command for each rectangle is given. You will record your predictions in the first column, then use the Scale tool to check your work. See if the commands in the Command Center match your predicted commands. If they match, put a check mark in the box by the word *Predict*. If they don't match, figure out what is wrong. You'll circle your errors, and make the needed corrections.

But first I'm going to demonstrate the other computer activity you'll be doing.

On Computer

Similar Houses

Demonstrate choosing **Change Activity** from the **File** menu and open the Similar Houses activity, which uses a five-sided "house" figure to help students further understand similarity.

When you open the activity, a procedure for drawing a house appears in the Command Center. Click **[OK]** or press **<return>** and the turtle draws the house in the Drawing window.

Direct attention to Student Sheet 24 and explain that as they did with Similar Rectangles, students are going to predict and check procedures that draw houses that are similar to this house.

Note: In the next part of the demonstration, enter your own commands rather than students', so that you can include an intentional error.

The first similar house is supposed to have sides that are double the length of the original house. I'm going to predict what the fd commands should be.

Enter these new fd commands in the Command Center, replacing the old commands:

```
fd 16
fd 20
fd 20
fd 16
fd 30
```

(The intentional error is the last command, fd 30 instead of fd 32.) Run the new procedure. Because the last fd command is too short, the house will not quite close, indicating that it is not similar to the original house.

When the error is small like this, the turtle may cover the gap. Using the command ht (hide turtle) will make the gap clear. The turtle also disappears and the gap becomes clearer when you use the Scale tool to check. After making a visual check, students should still check their numbers as follows:

With the Scale tool, scale down so the first command is fd 8 again. Then compare the commands in the Command Center to the original procedure (either in the Teach window or on Student Sheet 24). Point out that in this example, the last fd command does not match the original procedure. Ask students how the last fd command should be changed to make a double-size similar house.

Scale back up to a first command of fd 16, correct the last command to fd 32, and run the procedure to demonstrate the correct similar house.

Use the chart for Similar Houses the same way as the one for Similar Rectangles. Write your predicted commands. Then run them. Is your house similar? Check with the Scale tool. If you have drawn a similar house, put a check mark at the top of that column. If your house is *not* similar, circle the commands that need to be changed and write your corrections next to them. Then try making the next house.

If students find that the Scale tool moves from their procedure down to 0 without showing a similar figure, this means their house is not similar. Suggest that they check their procedure for errors, make corrections in the Command Center, and try again.

Depending on your computer situation, you might want to send half the class to the computers with Student Sheets 23 and 24 now to get started while you introduce the off-computer activity to the other half. Or, you might introduce the off-computer activity to the whole class so that everyone knows the three tasks for these sessions.

Off Computer

Similarity Poster

Students work at their desks with Power Polygons. They start by building a nonregular polygon of their own choosing. Suggest that students use only 2–5 Power Polygons in their original shape. The **Teacher Note,** Building Larger Similar Figures (p. 110), shows some examples.

Then students use additional Power Polygons to make larger shapes that are mathematically similar to their original shape. They should make one with each side twice as long, and another with each side three times as long as the original. In addition, students make any other similar shapes they discover, either larger or smaller than the original shape.

To display their work as a poster, students trace around their original shape and the two (or more) similar shapes on a large sheet of paper. If students are able to use multiples of the whole original shape to make a similar shape, they draw a dark outline around the original shape anywhere it appears in the new shape. (As an example, see the shape made from Power Polygons A and E in the **Teacher Note,** Building Larger Similar Figures, p. 110.) However, many students will not have the whole original piece inside, which is fine.

Students color-code their work by making each piece with the same letter the same color. Those who start with large original shapes (like the boy shown below) may need an extra sheet of paper to show the third larger shape.

When students have completed their posters, they investigate whether their similar nonregular polygons follow the generalizations they discovered earlier about the way area changes as perimeter changes. They label side lengths and areas. What multiple of the original are the sides and area of each similar shape?

Suggest that students use pencil for this step, until they can check each other's numbers during the follow-up discussion, Sharing Conclusions About Similar Shapes.

Some of their shapes may be too difficult to compare to the original, but most students will be able to compare the shapes that have sides double and triple the original.

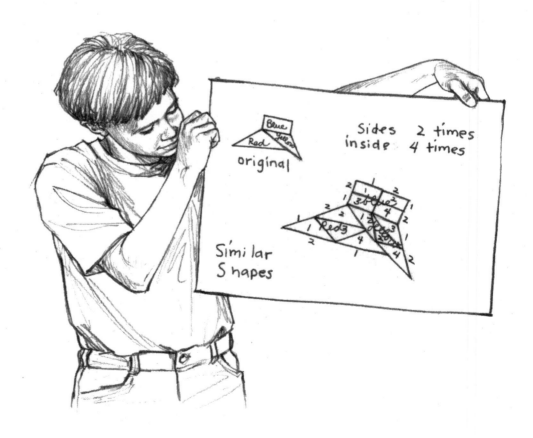

Classroom Management

After introducing the activities, assign half the class to work on the computers for the remainder of Session 5, while the other half works on the similarity poster. Switch groups during Session 6, allowing equal work time for each activity. Spend the last part of Session 6 discussing the work of these two sessions.

While Students Are Working off the Computer Observe the shapes students are making, and ask them to show you how they know they are similar. Remind them how to "eyeball" one shape held above another. Insist that they compare the angles of two figures to be sure they are the same, and measure the side lengths, being sure that they have all been multiplied by the same number so they are in the same proportion as the original.

While Students Are Working on the Computer If students have difficulty with the rectangles on Student Sheet 23 for which the first fd commands are fd 16 and fd 24, encourage them to think about the relationship between the side lengths in the original rectangle—10 and 15. See the **Teacher Note,** Writing Procedures for Similar Figures (p. 109) for more on students' ideas about and problems with similarity.

Ask students to describe their strategy for drawing similar figures. If students continually use addition, or add more turtle steps to the final forward command to close the house, remind them that they need to have a consistent strategy for changing all the lengths in the same way.

If students continue to have difficulty writing commands for similar houses, suggest that they let the computer do one for them, to see how it works. They use the Scale tool to stretch the figure until the first forward command matches the one they are working on (see How to Use the Scale Tool for Help in the *Geo-Logo* Tutorial, p. 142). They can study the new commands given in the Command Center and compare them to the original commands to find a pattern that will help them draw the other similar houses.

If students seem to want to use the Scale tool to find the commands for each house, encourage them to try the writing assignment at the bottom of Student Sheet 24, explaining how they figured out the commands for the triple-size house. This can help them focus on other strategies.

If they have extra time on the computer, student pairs can challenge each other on the Similar Rectangles activity. One gives a length for the vertical pair of sides of the rectangle; the other tries to figure out what the length of the other pair of sides will be. Students use the Scale tool to check their answers. Students who are interested in this further exploration might want to show one or more decimal places for the side lengths. They can make this change by choosing Decimal Places under the Options menu. Note that, due to rounding, the numbers in the Command Center may not be precise when one or more decimal places are shown.

Activity

Sharing Conclusions About Similar Shapes

Toward the end of Session 6, students get together in groups of three or four to compare their posters. They explain how they made sure their shapes were similar. They check one another's work and help find how many of the original sides and areas fit in each similar shape. When they are sure of these numbers, they may write them on the poster with crayon or marker.

When posters are checked and completed, display them around the room, separated into two groups—those whose larger similar shapes are made by repeating the original shape, and those that are not.

Some of you were able to enlarge your shape by making multiple copies of the shape and arranging them to make a similar shape. Which shapes were possible to enlarge this way? Which shapes were not?

What strategy did you use to enlarge your shape if you didn't repeat the original shape?

Note: All three- and four-sided convex figures will tessellate, which means that you can use four of the original shape to get the next larger similar figure). Concave quadrilaterals (such as chevrons) and polygons with more sides will often grow only by making similar copies of each individual shape within the original shape, as shown in the **Teacher Note,** Building Larger Similar Figures (p. 110).

As students circulate to look at the posters, they may write questions or comments on stick-on notes and attach them to the posters. Allow time for students to answer the questions that come up and discuss their ideas.

Choosing Student Work to Save

As the unit ends, you may want to use one of the following options for creating a record of students' work on this unit.

■ Students look back through their folders or notebooks and write about what they learned in this unit, what they remember most, and what was hard or easy for them. Students might do this during their writing time.

■ Students select one or two pieces of their work as their best, and you also choose one or two pieces, to be saved in a portfolio for the year. Students can create a separate page with brief comments describing each piece of work.

■ You may want to send a selection of work home for families to see. Students write a cover letter, describing their work in this unit. This work should be returned if you are keeping year-long portfolios.

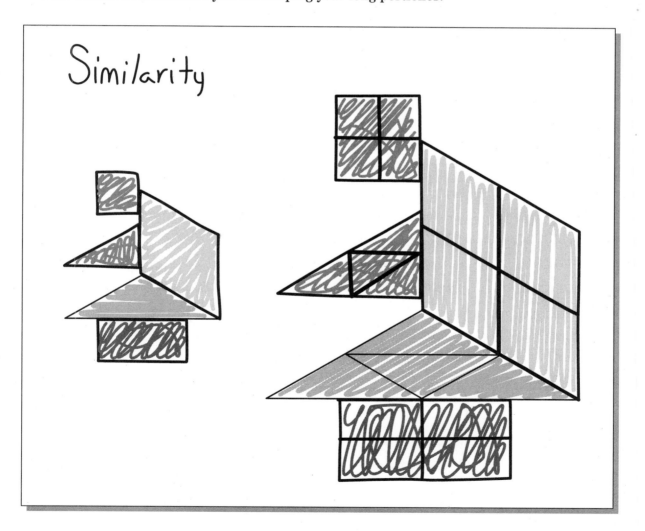

Sessions 5 and 6 Follow-Up

 **Homework**

Drawing More Similar Rectangles Send home 2–3 sheets of one-centimeter graph paper and Student Sheet 25, Drawing More Similar Rectangles. Students draw a rectangle on the graph paper and then draw another that is similar to it. They then challenge someone at home to draw yet another rectangle that is similar to the first two. Or, they might provide one side of the similar rectangle and then challenge the person to draw the rest, making it similar to the original. They can take turns challenging each other. Encourage students to experiment with wide and narrow rectangles, making a few that are similar to each.

 **Extension**

Growth of Perimeter and Area in Similar Shapes Students start by drawing a 2 × 3 rectangle on one-centimeter graph paper, then make many rectangles that are similar to the original. They may use rulers or any other tools they find helpful. They tell in writing how they know all the rectangles are similar, and record under each shape the dimensions, the perimeter, and the area of the shape.

This will be the first time students have compared whole perimeters of similar rectangles. They should figure out that if all the side lengths are multiplied by a number, then the whole perimeter is multiplied by that number. Relationships will be clearer if students make a table of the perimeters and areas, starting with the smallest.

Rectangle	Perimeter	Area
2 × 3	10 (original)	6 (original)
1 × 1.5	5 ($\frac{1}{2}$ original)	1.5 ($\frac{1}{4}$ original)
4 × 6	20 (2 × original)	24 (4 × original)

2×3   1×1.5   4×6

Writing Procedures for Similar Figures

This investigation develops students' ideas about similar figures, ratio, and proportion. Students write *Geo-Logo* procedures that draw figures that are similar to a given figure.

Developing strategies for building and drawing similar shapes is often difficult for students. These are some common problems they might have:

■ Students may confuse the mathematical and general meanings of words such as *similarity* and *proportion*.

■ Students will often begin by using an additive strategy to build similar figures. ("To get from fd 8 to fd 16 they added 8, so I'm going to add 8 to all the forward commands.") This mistake and its result is shown in the diagram below.

When their figure does not close, students will often try to add a few more turtle steps to the final forward command to close the figure because they do not recognize the importance of having a consistent strategy for changing the measures of the figures. These students need to check for themselves and see that the house they have made is not similar to the original house. After drawing a new house, they use the Scale tool to stretch or shrink that house to see if its numbers match those of the original.

Seeing the relationship between similar figures as multiplicative rather than additive will take many students time on the computer to experiment and get feedback about their ideas. Ultimately, students should adopt a consistent multiplicative strategy and become aware that this is necessary to make a similar shape.

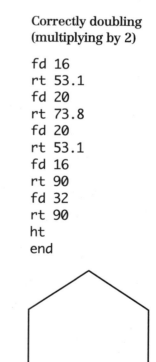

Original	Adding 8	Correctly doubling (multiplying by 2)
fd 8	fd 16	fd 16
rt 53.1	rt 53.1	rt 53.1
fd 10	fd 18	fd 20
rt 73.8	rt 73.8	rt 73.8
fd 10	fd 18	fd 20
rt 53.1	rt 53.1	rt 53.1
fd 8	fd 16	fd 16
rt 90	rt 90	rt 90
fd 16	fd 24	fd 32
rt 90	rt 90	rt 90
ht	ht	ht
end	end	end

Shapes with Three and Four Sides All three- and four-sided convex shapes can be made into larger similar figures by tessellating the original shape. For example, given a shape made with Power Polygons A and E, students could make a similar larger shape with four copies of the original.

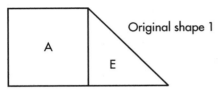

Original shape 1

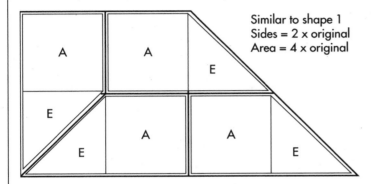

Similar to shape 1
Sides = 2 x original
Area = 4 x original

Students may also find many ways to use *different* pieces to make similar shapes for A and E. For example, they might make a smaller similar shape with B and F, and one (same-size) with E and F.

Similar to shape 1
Sides = $\frac{1}{2}$ original
Area = $\frac{1}{4}$ original

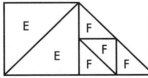

Similar to shape 1
Sides = original
Area = original

Shapes with More Than Four Sides Shapes with more than four sides often require the use of different pieces to create similar shapes. Consider this original:

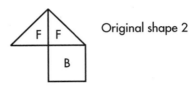

Original shape 2

In order to create a similar shape using only B and F pieces, students could use four of each to construct the similar shape with four times the area:

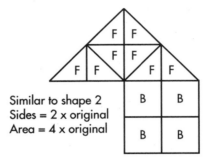

Similar to shape 2
Sides = 2 x original
Area = 4 x original

Using different pieces, students could construct a variety of similar shapes, as shown in the following diagrams.

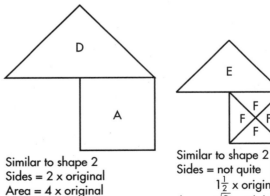

Similar to shape 2
Sides = 2 x original
Area = 4 x original

Similar to shape 2
Sides = not quite
     $1\frac{1}{2}$ x original
Area = $\sqrt{2}$ x original

Continued on next page

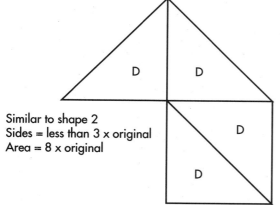

Similar to shape 2
Sides = less than 3 x original
Area = 8 x original

Starting with the original shape below, students could *not* use four figures of the same shape to construct a similar figure. Instead, they could use four of each individual shape (D, G, J, and A) to make larger, similar figures that would go together to make a shape similar to the original and four times as large. This is the only way to make similar shapes with more than four sides using the original shapes.

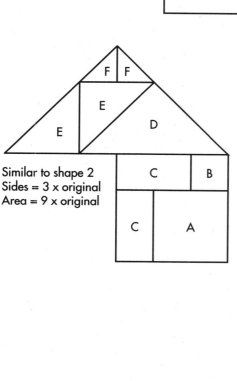

Similar to shape 2
Sides = 3 x original
Area = 9 x original

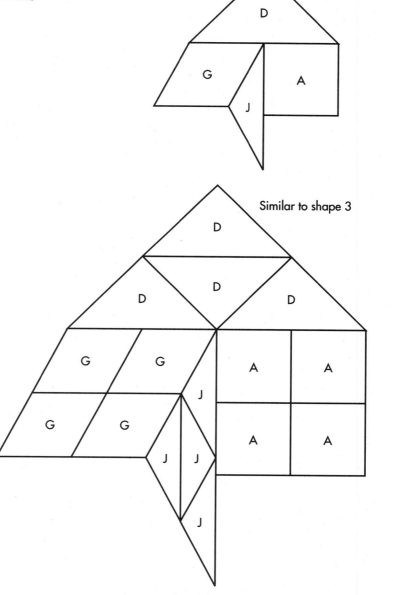

Shape 3

Similar to shape 3

Multiple and Factor Bingo

Basic Activity

Multiple and Factor Bingo can be played either as a whole class, with a partner, or in a small group. Students can play independently, but it's more fun to play with other people. The object of the game is to mark five numbers in a row, either across, up and down, or diagonally. The numbers that can be marked are determined by drawing cards.

These versions of the traditional Bingo game are designed to give students practice finding factors and multiples of numbers. Whether players are skip counting, using multiplication pairs, or dividing, this activity fosters numerical reasoning about the relationships among factors and their multiples. Students focus on:

- relating factors to their multiples
- becoming familiar with multiplication patterns
- developing number sense about multiplication and division relationships

Materials

- 100 (or 300) Chart for Multiple Bingo (1 per player for each game)
- Multiple Bingo Cards (1 deck per playing group)
- Multiplication Table for Factor Bingo (1 per player for each game)
- Factor Bingo Cards (1 deck per playing group)
- Crayon or marker (1 per player)
- Calculators (optional)

Procedure

Step 1. Gather the materials. For Multiple Bingo, each player has a 100 or 300 chart and a crayon or marker. Each playing group has a deck of Multiple Bingo cards in a pile face down in the middle of the table.

For Factor Bingo, each player has a Multiplication Table and a crayon or marker. Each playing group has a deck of Factor Bingo cards in a pile face down in the middle of the table.

Step 2. Draw a card from the face down pile. Players take turns turning over a card for the group.

Step 3. Choose a number to mark. For Multiple Bingo, every player marks one number on the 100 chart that is a *multiple* of the number on the card drawn. Players write the original number in a corner of the square for checking later. For example, if someone turns over a 5 card, players could mark any one of the numbers 5, 10, 15, 20, 25, and so forth. In the corner of the chosen square, the player would write 5. (With the 300 chart, players will simply have more options.)

Factor Bingo works the same way, except every player marks one number on the Multiplication Table that is a *factor* of the number on the card drawn. For example, if someone turns over a 100 card, players could mark any one of the numbers 1, 2, 4, 5, 10, 20, 25, 50, or 100. In the corner of the chosen square, the player would write 100.

In either game, if a Wild Card is drawn, the player who turned it over can decide on the number to be used. The best strategy is to choose a number that helps the player's own game but doesn't help the other players. For example, in Multiple Bingo, the most useful number to pick is often a prime number. For example, a player might pick 23 to fill in a gap on his or her card, knowing that other players' choices would be limited to 23, 46, 69, or 92.

Step 4. Repeat the process until there is a winner. The game continues until a player marks five numbers in a row for a Bingo. The remaining players can choose to continue until they also mark five in a row.

Continued on next page

Variations

Whole-Class Game This game can be played by a whole class. A leader, the teacher or a student, draws the cards. If a Wild Card is drawn, the leader calls on a player to choose a number for the group to use. The object of the game remains the same. Play could continue until every player has covered five in a row. When the class plays as a large group, there is the option for students who are new to the game to collaborate with other students.

Limiting the Cards For an easier version of Multiple Bingo, use only the 2, 3, 4, and 5 cards and a few Wild Cards. As students become comfortable with additional multiples, add cards to the game.

For an easier version of Factor Bingo, use only the top two rows of Factor Bingo Cards (100, 180, 200, 60, 98, 32, 72, and 150) and a few Wild Cards.

Limiting the 100 Chart When students first play Multiple Bingo, they will tend to use only "easy" numbers, especially the single-digit numbers and multiples of 10. Here are some ways to encourage them to use more difficult numbers:

- Block out the top row and right-hand column of the 100 chart; these numbers may not be marked.

- Establish a rule that players must start with a number near the middle of the chart.

- Give bonus points for a win that is on a diagonal. This may encourage students to notice the nines and elevens tables on the two main diagonals.

Special Notes

Using Resources Encourage students to use 100 charts and calculators as a way of determining suitable factors and multiples.

Reusing the Charts Students can use a contrasting color to play another game on the same 100 chart or Multiplication Table. Some teachers have laminated a set that can be wiped off after each game; students will need to use either crayons or white-board markers with these laminated charts.

The following activities will help ensure that this unit is comprehensible to students who are acquiring English as a second language. The suggested approach is based on *The Natural Approach: Language Acquisition in the Classroom* by Stephen D. Krashen and Tracy D. Terrell (Alemany Press, 1983). The intent is for second-language learners to acquire new vocabulary in an active, meaningful context.

Note that *acquiring* a word is different from *learning* a word. Depending on their level of proficiency, students may be able to comprehend a word upon hearing it during an investigation, without being able to say it. Other students may be able to use the word orally, but not read or write it. The goal is to help students naturally acquire targeted vocabulary at their present level of proficiency.

We suggest using these activities just before the related investigations. The activities can also be led by English-proficient students.

Investigation 1

shape, sides, triangle, quadrilateral, impossible

1. Hold up a piece of paper, outline its shape with your finger, and count the *sides.* Explain that this *shape* has four sides, that it is a *quadrilateral.*

2. Next, cut the paper from corner to corner to make a triangle. Point out that the shape has changed, and count the sides of the new shape.

3. Challenge a student to use the scissors to cut the paper into yet another shape and to count the sides.

4. Ask other students to continue changing the shape of the paper, making shapes with different numbers of sides. Each time, count the sides.

5. Finally, challenge students to cut the paper into a shape with just *one* side. When they realize this can't be done, say that this is an *impossible* task. Ask them to name other

things that are impossible to do.

Investigation 3

doubled, tripled, original

1. Write three headings on the board: *Original, Double,* and *Triple.*

2. Place one penny on a table, and write "1 penny" under the heading *Original.*

3. Place another penny near the first one, and tell students that you have *doubled* the *original* amount. Under *Double,* write "2 pennies."

4. Now place a third penny on the table, telling students that you have now *tripled* the original amount. Make the appropriate entry on the board. You may want to add designations "× 2" and "× 3" by the *Double* and *Triple* headings.

5. Ask about what you just did, posing questions that require one-word responses.

   How many pennies were in the original amount?

   How many pennies were there when I doubled the amount? How many when I tripled the amount?

6. Repeat the activity, varying the number of pennies used for the original amount. Continue to ask questions that require students to identify the number needed to double and triple each original amount.

   You can use coins other than pennies or any other object for this activity, as long as the objects are identical to one another.

Teacher Tutorial

Contents

Overview

The units in *Investigations in Number, Data, and Space* ask teachers to think in new ways about mathematics and how students best learn math. Units such as *Picturing Polygons* add another challenge for teachers—to think about how computers might support and enhance mathematical learning. This Tutorial can help you learn what the computer component is, how it works, and how it is designed to be used in the unit.

The Tutorial is written for you as an adult learner, as a mathematical explorer, as an educational researcher, as a curriculum designer, and finally—putting all these together—as a classroom teacher. Although it includes parallel (and in some cases the same) explorations as the unit, it is not intended as a walk-through of the student activities. Rather, it is meant to provide experience using the computer program *Geo-Logo*™ and to familiarize you with some of the mathematical thinking in the unit.

The Tutorial is organized in sessions parallel to the unit. Included in each session are detailed step-by-step instructions for how to use the computer and the *Geo-Logo* program, along with suggestions for exploring more deeply. The later parts of the Tutorial include more detail about each component of *Geo-Logo* and can be used for reference. There is also detailed help available in the *Geo-Logo* program itself.

In *Picturing Polygons,* students use *Geo-Logo* for mathematical, particularly geometric, exploration. With *Geo-Logo,* students are able to construct paths and geometric shapes in addition to observing them. Since one of the best ways to learn something is to teach it, *Geo-Logo* uses the metaphor of "teaching the turtle" how to move, turn, and draw. Writing a list of instructions for how to construct a shape encourages students to think carefully about geometric properties and to use geometry-oriented language.

Teachers new to using computers and *Geo-Logo* can follow the detailed step-by-step instructions. Those with more experience might not need to read each step. As is true with learning any new approach or tool, you will test out hypotheses, make mistakes, be temporarily stumped, go down wrong paths, and so on. This is all part of learning but may be doubly frustrating because you are dealing with computers. It might be helpful to work through the Tutorial and the unit in parallel with another teacher. If you get particularly frustrated, ask for help from the school computer coordinator or another teacher more familiar with using computers. It is not necessary to complete the Tutorial before beginning to teach the unit. You can work through it in parts, as you prepare for the investigations in the unit.

Although the Tutorial will help prepare you for teaching the unit, you will learn most about *Geo-Logo* and how it supports the unit as you work side by side with your students.

Note to teachers: These directions assume that *Geo-Logo* Picturing Polygons has been installed on the hard disk of your computer. If not, see How to Install *Geo-Logo* on Your Computer, p. 156.

1. Turn on the computer by doing the following:

   a. If you are using an electrical power surge protector, switch to the ON position.

   b. Switch the computer (and the monitor, if separate) to the ON position.

   c. Wait until the desktop or workspace appears.

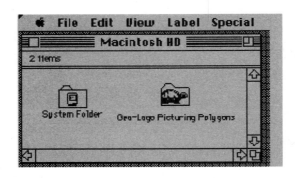

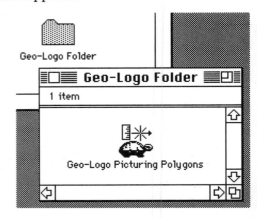

2. Open *Geo-Logo* by doing the following:

   a. Double-click on the *Geo-Logo* folder icon if it is not already open. To double-click, click twice in rapid succession without moving the pointer.

   b. Double-click on the *Geo-Logo* icon in this folder.

   c. Wait until the *Geo-Logo* opening screen appears. Single-click on this screen when the "click" message appears.

Investigations in Number, Data, and Space®
Picturing Polygons
Geo-Logo™
Click on this window to continue.
Authors & Programmers:
Douglas H. Clements & Julie Sarama
In collaboration with Michael T. Battista
© D. H. Clements, 1993; Logo core is © LCSI, 1993
Activities © 1996 Dale Seymour Publications®

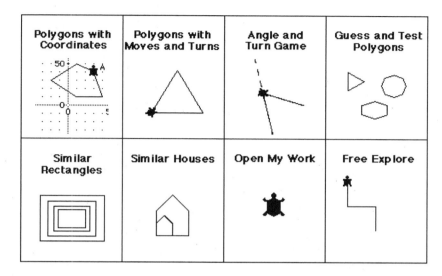

How to Start an Activity

Start an activity by doing the following:

1. Single-click on the **[Polygons with Coordinates]** activity (or on any activity you want).

   When you choose an activity, the Tool bar, Command Center, Drawing window, and Teach window for that activity appear on the screen.

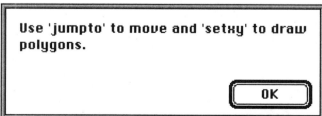

A dialogue box appears with directions.

2. Click on **[OK]** or press the **<return>** key to close the dialogue box. You are ready to begin working in Polygons with Coordinates. Should you need them, trouble-shooting notes are included at the end of the tutorial.

How to Make Polygons with Coordinates

The turtle in *Geo-Logo* is a robot that follows your commands. Throughout the tutorial, you will command the turtle to turn and move along various paths and to jump to given locations with coordinate commands. For Polygons with Coordinates, you use two coordinate commands, `jumpto` and `setxy`, to draw polygons.

1. Enter commands in the Command Center to make the turtle go from its starting position, coordinates (0, 0), to the starting position for your picture. Try your own commands or follow these steps (`jt` stands for `jumpto`; the full command can also be used):

   a. Type: `jt [-20 -40]`

   That's twelve keystrokes: j, t, space, left square bracket, hyphen, 2, 0, space, hyphen, 4, 0, right square bracket.

   You can use the **<delete>** key to make changes, if needed. If the text you type does not appear in the Command Center, click somewhere in the Command Center to make it the active window—the one that receives text.

   b. Press **<return>**.

   The turtle jumps to that coordinate position without drawing. Notice that the distance from each dot on the coordinate to the next—horizontally or vertically—represents 10 units of distance.

   c. Type: `setxy [0 -10]`

   d. Press **<return>**.

   Typing a command and pressing **<return>** to run it is called *entering* a command. Notice that the turtle draws a line segment when you run the `setxy` command.

   Notice that the *Geo-Logo* commands `setxy` and `jumpto` (or `jt`) use square brackets and no comma between numbers for the coordinates of a point, instead of parentheses and commas as used in traditional mathematical notation. In *Geo-Logo*, parentheses and commas have been reserved for different meanings.

   e. Continue to enter these commands to draw a polygon.

   ```
   setxy [30 -40]
   setxy [40 -30]
   setxy [10 0]
   setxy [40 20]
   setxy [-30 30]
   setxy [-20 -40]
   ```

Remember to think of the turtle as a robot that follows certain *Geo-Logo* commands. If it does not understand a command, it will write a message in a dialogue box. See *Geo-Logo* Messages, p. 154.

To edit (change) your commands, use the **<delete>** key to back over and erase text. Type new text and press **<return>**. Another way to edit commands is to use the mouse to select words or blocks of text by dragging (pressing and holding down the mouse button as you move the mouse) over the text. Then press **<delete>** and type new text. You can also use the arrow keys to move the text cursor (the blinking line that shows where the next type will appear) from place to place in the Command Center.

A third way to change commands is by using the erase options in *Geo-Logo* tools. Above the Command Center is the Tool bar, a row of icons for tools available in this software. To erase commands, click on one of these:

Erase One (erases only the last command listed)

Erase All (erases everything in the Command Center)

Each time you change a command and press **<return>**, the turtle (in the Drawing window) reruns the commands.

2. Check to see if your procedure draws the polygon you wanted.

   If not, make changes in the Command Center. Move up or down with the mouse or arrow keys, use the **<delete>** key to erase, and type in your change. After you finish all your changes, press **<return>** and you will see the effect of these changes in the Drawing window.

3. Teach the turtle a new procedure to draw your polygon by following these steps:

   a. Click on the Teach tool.

      When you click on the Teach tool, you define the list of commands in the Command Center as a procedure. The computer will ask you to type a name for the procedure.

   b. Type a one-word name for this procedure, such as *arrow.* You can use letters or numbers.

   c. Click **[OK]** or press **<return>**.

The procedure appears in the Teach window. It is defined by the name you give it. Notice that the computer adds a first and last line to your commands when you define them as a procedure. The first line is to arrow (or whatever name you chose), and the last line is end.

Notice that the drawing window clears in preparation for your next entry.

You have taught the turtle to draw a particular polygon. If you want the turtle to draw it again, go to Step 4.

4. Run your procedure by following these steps:

   a. Type your procedure's name, for example, arrow, in the Command Center.

      If needed, move the text cursor (the blinking vertical line that shows where any typed text will go) into the Command Center by clicking the mouse in that window.

   b. Press **<return>**.

If you want to make changes to your procedure, you may edit any commands in the Teach window. Click your cursor in that window, move up or down with the mouse or arrow keys, use the **<delete>** key to erase, and type in your changes. When you click out of that window, the turtle will rerun your procedure to show your changes.

5. Choose the **Help** menu and explore what assistance is available when you choose **Windows, Vocabulary** (Commands), **Tools, Directions,** or **Hints.**

6. Click on the Erase All tool to clear the Command Center and Drawing windows in preparation for drawing a new polygon.

   Notice that any procedures you have defined are still available in the Teach window.

How to Save Your Work

When you turn off the computer or start a new activity, the computer's memory is cleared of all commands and procedures to make room for new ones. To avoid losing your work, you can save it on a disk before the computer memory is cleared. Once your work has been saved on a disk, you can open it again to show it to someone or to work on it some more.

1. Choose **Save My Work** from the **File** menu:

   a. Move the mouse pointer over the word **File** in the menu title bar along the top of the screen.

   b. Press and hold the mouse button until the menu items appear.

   c. Continue to press the mouse button as you drag the pointer down and select **Save My Work**.

   d. Release the mouse button.

The notation ⌘S indicates that you can use a keyboard command instead of the **File** menu. In this case, hold down the ⌘ key and type <S>. Entering ⌘S and choosing **Save My Work** from the **File** menu give the same result.

The first time you save your work, a dialogue box like this one will appear asking for a name.

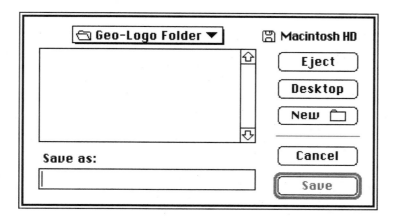

2. Type a name for your work.

   You can choose any name. However, useful names include information that helps you find the work again, such as your initials, the activity, and the date. Spaces can be used in the titles of saved work. An example would be:

   JH Poly with Coord 9/22

   You can also choose a different location for your work. The bar with the name **Geo-Logo Folder** tells where the work will be saved. It also is a pop-up menu that can be used to choose a different folder into which you can save your work.

3. Click on **[Save]**.

   Notice that the name of your work now appears in the title bar of the Drawing window.

   When you save your work this way, a copy is stored on the computer disk. You can stop using the computer for a while or come back to this work again at another time.

How to Stop an Activity

1. To stop working on this activity:

   Choose **Close My Work** from the **File** menu. If you have not already saved your work, or if you have made any changes since you last saved it, a dialogue box will appear asking whether you wish to save it.

   Notice that once you close your work, the computer is ready to start new work on this activity. If others will be doing the same activity, you may want to leave the screen like this instead of quitting *Geo-Logo* and shutting down the computer.

2. To stop using *Geo-Logo*:

   Choose **Quit** from the **File** menu.

3. To stop using the computer:

   a. Follow the usual procedure to shut down and turn off your computer.

How to Continue with Saved Work

If needed, start your computer, open *Geo-Logo*, and select **[Polygons with Coordinates]** (or whatever activity you wish) using steps explained earlier.

1. To continue with your previous work:

   a. Choose **Open My Work** from the **File** menu.

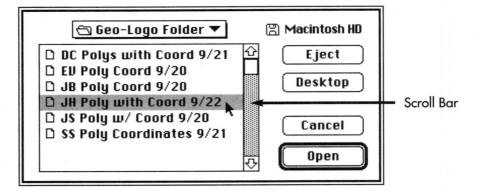

Scroll Bar

   If your work has been saved on a different disk, insert that disk, click on **[Desktop]**, and choose that disk from the menu.

   b. Select work by clicking on its title in the list. You may need to scroll up or down the list to find your title; click on the up or down arrows in the scroll bar to do that.

   c. Click on **[Open]** (or double-click on the title).

Remember that assistance is available from the **Help** menu at any time: Choose **Windows, Vocabulary, Tools, Directions,** or **Hints** to get information.

Tools   The full *Geo-Logo* Tool bar, which you will see in Free Explore, looks like this:

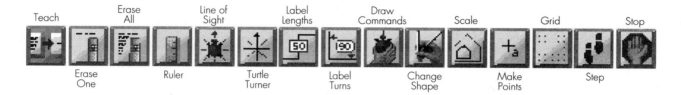

For a particular activity, only the tools most commonly used are displayed. Other tools will be used in the other activities on this disk. Following are the tools you are most likely to use in Polygons with Coordinates.

	Teach	Teaches the turtle the commands in the Command Center as a new procedure
	Erase One	Erases the last command.
	Erase All	Erases all the commands in the Command Center.
	Label Lengths	Shows length of line segments in turtle steps on the Drawing window
	Step	Walks you through one command at a time, either in the Command Center or in a procedure in the Teach window, to help find errors and to edit. The cursor turns into "walking feet." Click in the Drawing window to see each step. Click in the Command Center or the Teach window to stop "stepping." This allows you to change a command.
	Stop	Stops commands that are running.

For more details about these and other tools, see Tools, p. 149.

Notes

Choose **Show Notes** from the **Windows** menu to record thoughts and observations about your work. (see the example below). To close Notes, choose **Hide Notes** from the **Windows** menu or click in the close box on the left of the Notes title bar.

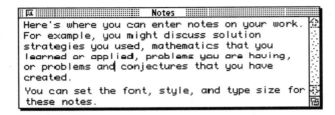

Printing

You can print all your work, including the picture and list of commands and procedures. Choose **Print** from the **File** menu.

If the Drawing window does not print completely, try selecting the Color/Grayscale option for printing.

You can also print just a single window. For example, if you want only a list of your commands or only a copy of your picture, click into that window and choose **Print Window** from the **File** menu.

Text Size

You can enlarge all the text in the Command Center and Teach window for easier viewing. Select **All Large** from the **Font** menu. Select **All Small** to change back to the normal font size.

Making Shapes That Follow Rules

For this activity, you use *Geo-Logo*'s coordinate commands to make shapes with given properties, specified on Student Sheet 9, Can You Make These Triangles? and Student Sheet 10, Can You Make These Quadrilaterals? As you read this section of the tutorial, it is useful to have a copy of the tools available in *Geo-Logo*, as shown on p. 149. Making a copy of this page to use as you read the tutorial also might be helpful.

If you need to, restart *Geo-Logo* and choose **[Polygons with Coordinates]** on the *Geo-Logo* activities screen. When the dialogue box appears, click **[OK]**.

1. Enter commands to draw tri1 from Student Sheet 9, a triangle with one right angle.

   a. Enter: `jt [0 35]`

   This positions the turtle at (0, 35). To create a right angle, we need perpendicular sides.

   b. Continue to enter these commands to draw a right angle.

   `setxy [0 10]`
   `setxy [40 10]`

   Because we first changed only the *y*-coordinate, we know we made a vertical side; similarly, we then changed only the *x*-coordinate, making a horizontal side. So, we must have made a right angle.

   c. Enter: `setxy [0 35]`

   Returning to the starting point creates a closed shape.

2. Use the Teach tool to define your procedure.

3. Run the new procedure by entering its name in the Command Center.

The triangle called tri2 on Student Sheet 9, with two right angles, is impossible to make. If two angles of a figure are right angles, it has two parallel sides, so they cannot meet to form a triangle.

The triangle tri3 is difficult to make with `setxy`, unless you know higher-level mathematics and use decimals. (In the next section, we will use other commands that make this task easier.) One solution is as follows:

`setxy [20 0]`
`setxy [10  17.32055]`
`setxy [0 0]`

Use the Label Lengths tool to determine if the lengths are equal. Click on it again to remove the labels.

Try making tri4 and tri5 yourself.

You can make quad1 on Student Sheet 10, with four right angles and two pairs of equal and parallel sides, with a strategy similar to the one you used for the right angle. Try it.

The quad2, with no right angles but four equal sides, may be more difficult for some students. One good way to approach this task is shown below.

4. Enter commands to draw quad2.

   a. Enter:

```
jumpto [0 0]
setxy [-10 20]
setxy [0 40]
setxy [10 20]
setxy [0 0]
```

   b. Use the Label Lengths tool to determine if the lengths are equal.

     Click on this tool to label the lengths of all line segments. What are their lengths? If they are not equal, your procedure needs correction. Click on the Label Lengths tool again to turn the labels off.

5. Use the Teach tool to define your procedure.

6. Run the new procedure by entering its name in the Command Center.

The quad3, with four right angles and no equal sides, is impossible to make. The right angles force a rectangle, and so also force equal sides.

If you like, try quad4 and quad5 yourself.

Polygons with Moves and Turns

1. If you are already in *Geo-Logo*, choose **Change Activity** from the **File** menu.

   A dialogue box may appear asking whether you wish to save your work. If you do, see How to Save Your Work, p. 122. If not, click **[Don't Save]**.

2. Single-click on the new activity—this time, **[Polygons with Moves and Turns]**—on the *Geo-Logo* activities screen.

How to Choose a New Activity

As Polygons with Moves and Turns opens, a dialogue box appears with instructions. Click **[OK]**.

```
Use 'fd', 'bk', 'rt', 'lt', and 'repeat' to draw
polygons. (Don't forget about 'jumpto'.)

                                    OK
```

How to Draw with Moves and Turns

In this group of activities, you command the turtle to turn and move along various paths. This approach emphasizes turns and angles. It also allows you to define procedures that can be drawn at any location and heading.

1. Enter commands in the Command Center to make the turtle draw a square:

   a. Type: `fd 20`

   b. Press **<return>**.

   The turtle moves 20 mm forward, drawing a line segment. Note that no brackets are needed, for 20 is a distance to move, not a pair of coordinates.

   c. Type: `rt 90`

   d. Press **<return>**.

   The turtle turns 90° on the center of its belly.

   e. Continue to enter these commands to draw a square.

```
fd 20
rt 90
fd 20
rt 90
fd 20
rt 90
```

Notice that the square looks complete after you enter the last fd 20. However, it's always a good idea to get the turtle back to its exact starting position and heading, so we enter the final turn rt 90.

2. Click on the Erase All tool to erase these commands.

There's a bit of repetition in the commands to draw a square, so this is a good time to introduce a new command, repeat.

How to Use the Repeat Command

The repeat command repeats, a specified number of times, any commands we list in square brackets. Thus, to make the square you made above, you would enter the command repeat 4 [fd 20 rt 90].

1. Make shapes with the repeat command.
   a. Type: repeat 2 [fd 20 rt 90 fd 40 rt 90]
   b. Press <return>.

   The turtle repeats the commands in the list (the bracketed commands) two times, drawing a rectangle.
   c. Click on the Erase All tool.
   d. Enter repeat 4 [fd 30 rt 90] to draw a square.
2. Use the Teach tool to define your procedure.
3. Run the new procedure by entering its name in the Command Center.

How to Make Polygons with Moves and Turns

In this activity, you command the turtle to turn and move along various paths to draw polygons with specific properties.

1. Enter commands in the Command Center to make the turtle draw tri3. (Many different procedures are possible.)
   a. Enter: repeat 3 [fd 40 rt 120]
   b. Use the Label Lengths tool to label the sides and see whether they are equal. They are, of course.
   c. Use the Label Turns tool to label the turns and see whether they are equal.

   This tool is especially helpful to students, who can sometimes become confused as to what is being measured, the turn or the interior angle.

The turns (exterior angles) labeled by the Label Turns tool are 120°. What are the internal angle measures? In an equilateral triangle, they are all 60° (180° – 120°).

When the turtle turns 45° and goes forward, the angle that is created measures 135° (180° – 45°). The *larger* the turtle turn, the smaller the angle created.

2. Use the Teach tool to define your procedure.

3. Run the new procedure by entering its name in the Command Center.

Making other triangles and quadrilaterals can be difficult with moves and turns if you're not sure of a measure or two.

1. Begin making tri1, a triangle with one right angle.

a. Enter these commands:

```
fd 34
rt 90
fd 34
```

Because this is an isosceles triangle, you might already know that the next turn will be 135, creating a 45° angle. But let's measure it anyway.

2. Measure the turn with the Turtle Turner tool.

a. Click on the Turtle Turner tool. One arrowhead shows the turtle's heading.

b. Move the other arrowhead with the mouse.

c. Click on the point toward which you want the turtle to turn, to "freeze" this arrowhead and show a turn command.

A dialogue box appears:

3. Enter rt 135 in the Command Center.

How to Use the Turtle Turner and the Ruler

An alternative to the Turtle Turner is the Line of Sight tool. To get a quick visual sense of the turn you want, click and hold on the Line of Sight tool; you can count the rays (each ray represents 30°) to determine the measure of the turn.

The turtle has made just the turn we wanted. It is heading toward its initial position to make a closed shape. However, what length is the last side?

4. Measure lengths and distances with the Ruler tool.

   a. Click on the Ruler tool. One end of the ruler is at the turtle's position.

   b. Move the other end of the ruler with the mouse.

   c. Click to freeze the ruler and show the length. Note that the mouse has a limited accuracy, so it is sometimes difficult to get an exact measure. Use the Ruler tool to get as close as you can, then adjust the measure if necessary.

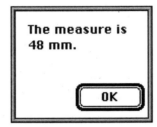

5. Enter commands to complete the right triangle.

   a. Enter these commands:

   ```
   fd 48
   rt 135
   ```

6. Use the Teach tool to define your procedure.

7. Run the new procedure by entering its name in the Command Center.

How to Begin Drawing at a Different Location

If you wish to begin a drawing at a location other than the center of the screen, you can use the jumpto command.

1. If you wish, click on the Grid tool to help you find an appropriate location.

2. Enter a jumpto command to position the turtle. For example:

   ```
   jumpto [-40 -30]
   ```

3. Enter the name of a procedure you have defined previously, or use fd and rt commands to draw.

Another way to begin drawing at a different location is to put the pen up (enter pu) and use the fd, bk, rt, and lt commands to move without drawing. Enter the pen down (pd) command to start drawing again.

How to Draw in Color

If you want to draw with different colors, you can use the setc command.

1. Enter a setc red command to change the turtle's color to red.

The turtle will now draw with a "red pen."

2. Enter repeat 4 [fd 30 rt 90] to draw a red square.

3. Enter these commands to fill the square with yellow:

pu rt 45 fd 10 setc yellow fill bk 10 lt 45 pd setc black

4. Enter pr colors to print a list of all the available colors. They are also listed in the **Vocabulary** selection of the **Help** menu under setc. See also fill.

Comparing Moves and Turns to Coordinate Commands

Often in *Geo-Logo* activities, we use move and turn commands (such as fd and rt) for drawing, with no coordinate commands involved. Students often believe that the coordinate commands are easier, and sometimes they are. Help students see that the move and turn commands like fd and rt have a real advantage when you want to use a procedure several times in different positions and orientations. Also, the turn commands let you learn and think about angles in more interesting ways.

How to Play the Angle and Turn Game

Open *Geo-Logo* and choose **[Angle and Turn Game]**. If you are already in *Geo-Logo,* choose **Change Activity** from the **File** menu. A dialogue box may appear asking whether you want to save your work. If you do, see How to Save Your Work, p. 122. If not, click **[Don't Save]**.

Single-click on the activity [Angle and Turn Game] on the *Geo-Logo* activities screen. A dialogue box appears with directions. Click **[OK]**.

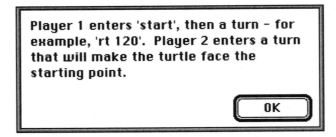

Player 1 enters 'start', then a turn – for example, 'rt 120'. Player 2 enters a turn that will make the turtle face the starting point.

OK

This two-player game requires students to integrate the measure of the turtle's turn and the measure of the angle created by the turn.

1. Enter start in the Command Center.

   The turtle goes forward from the starting point, then draws a dotted blue line to mark the initial heading.

2. Player 1 enters a turn command, such as rt 45.

   The turtle makes that turn, then the computer draws a green line segment to mark the turn, but the turtle does not move along the line.

3. Player 2 now enters a turn command. This command is supposed to turn the turtle to face back toward the starting point. This command should be the measure of the interior angle that is created by the initial turn and the forward movement before and after it. For this example, after Player 1 enters rt 45, the correct command for Player 2 is rt 135.

   The turtle makes that turn, then moves forward, drawing a red line segment. If the second turn is exactly right, the red line passes through the starting point. A dialogue box appears, specifying how much Player 2 is "off." The player who is off 0 degrees has made a perfect turn.

4. Continue to play a few more rounds of the game.

Obviously, the game is no longer interesting once you know that the second turn is 180 minus the first turn. The point of this activity is for students to discover and understand this strategy, and to understand the supplemental relationship (sums to 180) between any turn the turtle makes and the measure of the angle formed on the screen by the next fd command. (See the **Teacher Note,** The Rule of 180° on p. 68.)

Open *Geo-Logo* and choose **[Guess and Test Polygons]**. (Or, if you are already in *Geo-Logo,* choose **Change Activity** from the **File** menu, then single-click on the desired activity on the *Geo-Logo* activities screen.) A dialogue box appears with directions. Click **[OK]**.

How to Guess and Test Regular Polygons

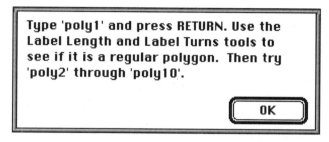

Type 'poly1' and press RETURN. Use the Label Length and Label Turns tools to see if it is a regular polygon. Then try 'poly2' through 'poly10'.

OK

This activity offers visual reinforcement in the concept of regular polygons.

1. Enter the first procedure, poly1, in the Command Center.

2. Guess whether that figure is a regular (equilateral and equiangular) polygon or not.

3. Use the Label Lengths tool to see if the sides are all the same length. Click on it again to turn it off.

4. Use the Label Turns tool to see if the turns (and therefore the angles in the figure) are all the same measure. Click on the tool again to turn it off.

5. Click on the Erase All tool.

6. Repeat steps 1 to 5, entering the procedures poly2, poly3, and so on, up to poly10.

How to Draw Regular Polygons

Open *Geo-Logo* and choose **[Polygons with Moves and Turns].** (Or, if you are already in *Geo-Logo,* choose **Change Activity** from the **File** menu, then single-click on the desired activity on the *Geo-Logo* activities screen.) A dialogue box appears with directions. Click **[OK]**.

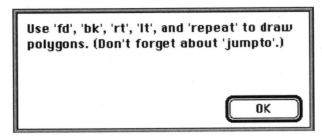

Use 'fd', 'bk', 'rt', 'lt', and 'repeat' to draw polygons. (Don't forget about 'jumpto'.)

OK

In this group of activities, you draw regular polygons with *Geo-Logo.* Here, coordinate commands would be virtually impossible for students. However, the move and turn commands are ideal. With a bit of searching for the patterns, the commands for any regular polygon can easily be found.

To begin, recall the commands that drew two regular polygons:

```
to equilateraltriangle
repeat 3 [fd 30 rt 120]
end
to square
repeat 4 [fd 30 rt 90]
end
```

In drawing these (and any regular polygon), the turtle turns a total of 360°: $3 \times 120 = 360$; $4 \times 90 = 360$. That makes sense: In traveling around the circumference of the figures, the turtle has made a whole turn of 360°. We can use that to figure out turns for any regular polygon.

1. Enter commands in the Command Center to make the turtle draw a pentagon:

   a. Enter: `repeat 5 [fd 20 rt 72]`

      You know the turn is 72 because $360 \div 5 = 72$.

2. Use the Teach tool to define your procedure as `pentagon`.

3. Run this procedure by entering `pentagon` in the Command Center.

4. Use the Label Lengths tool and the Label Turns tool to see that the measures of all the sides are equal, and the measures of all the turns are equal.

If *n* is the number of turns for a regular polygon, then the amount of turn at each vertex is 360/*n*. This is the Total Turtle Trip Theorem or the Rule of 360: For the turtle to complete a simple closed path (such as a polygon) and end up facing in the same direction as when it started, it must turn a total of 360°.

5. Enter commands in the Command Center to make the turtle draw a hexagon:

   a. Enter: `repeat 6 [fd 20 rt 60]`

      $360 \div 6 = 60$

6. Use the Teach tool to define your procedure as `hexagon`.

7. Run this procedure by entering `hexagon` in the Command Center.

8. Use the Label Lengths and Label Turns tools to see that the measures of all the sides are equal and the measure of all the turns are equal.

To make a seven-sided regular polygon, we need to find 360 ÷ 7, which requires decimals.

Geo-Logo can do that computation for us. Enter the command print 360 / 7 (with a space on each side of the slanted division sign) and press **<return>**. *Geo-Logo* will print the answer in the Print window (which opens automatically). Try it.

Some students may not readily see the patterns described above. In that case, they can approximate any regular polygon using two *Geo-Logo* tools:

 Draw Commands   Change Shape

The Draw Commands tool allows you to draw a polygon freehand, using the mouse to turn and move the turtle. As you draw, corresponding Logo commands appear automatically in the Command Center. You continuously create a turn (rt or lt) command, followed by a move (fd or bk) command.

1. Click on the Draw Commands tool.

   a. The cursor changes to cross hairs, the turtle continually turns to face it, and the corresponding turn command is dynamically updated in the Command Center. Click to freeze the turtle.

How to Use the Draw Commands and Change Shape Tools

Commands	Figure
lt 45 fd 10	
lt 45 fd 10 rt 40 fd 10.2 rt 33 fd 14.5 rt 61 fd 11.6 rt 60 fd 11.9 rt 30 fd 10.6 rt 41 fd 12.2 rt 49 fd 9.2	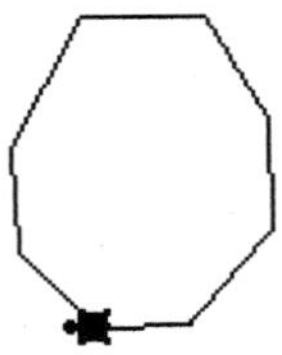
lt 45 fd 10 rt 40 fd 10.2 rt 43 fd 10 rt 51 fd 11.6 rt 34 fd 7.6 rt 54 fd 10.6 rt 41 fd 12.2 rt 49 fd 9.2	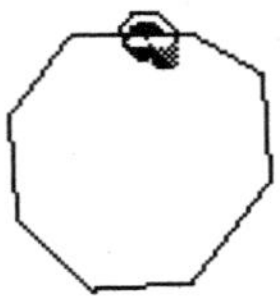
lt 45 fd 10 rt 45 fd 9.9 rt 45 fd 10 rt 45 fd 10.6 rt 45 fd 9 rt 45 fd 10.2 rt 45 fd 9.5 rt 45 fd 9.9	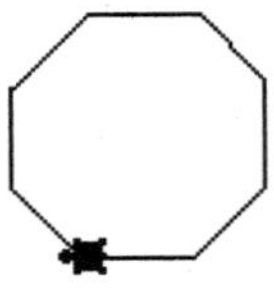

b. The cursor changes to a grabbing hand. Click on and drag the turtle forward or back. (To drag, hold the mouse button down while you move the mouse.) Release the mouse button to move the turtle to that position, as the corresponding fd or bk command appears in the Command Center.

c. Repeating this pattern can provide an approximation of the commands needed to draw the figure, in this case, a regular octagon.

d. Click in the Command Center to stop using the tool.

You can use another tool to adjust that first approximation. Let's say our approximation was as shown at the right.

2. Click on the Change Shape tool.

a. The turtle disappears. To move the top vertical line segment, click anywhere on the segment and drag it down to its new location.

b. Drag a corner to the new location. The commands change automatically. Release the mouse button when you are done.

c. Keep dragging corners or sides to get the best approximation you can.

d. Click in the Command Center to stop using the tool.

3. At this point, it makes sense to change the text in the Command Center directly, or look for a pattern and replace all these commands with:

repeat 8 [fd 10 rt 45]

Note: To use the Change Shape tool, the Command window must have a pattern of only fd and rt or fd and lt commands.

Open *Geo-Logo* and choose **[Similar Rectangles]**. (Or, if you are already in *Geo-Logo,* choose **Change Activity** from the **File** menu, then single-click on the desired activity on the *Geo-Logo* activities screen.) A dialogue box appears with directions. Click **[OK]**.

How to Make Similar Rectangles

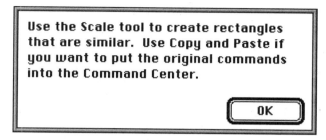

Use the Scale tool to create rectangles that are similar. Use Copy and Paste if you want to put the original commands into the Command Center.

OK

As the activity opens, the commands in the Command Center run and the turtle draws what is referred to as the "original" rectangle.

For this activity students are working with Student Sheet 23, Similar Rectangles, which asks them to draw four rectangles similar to the original one, working from a given side length.

1. Use the Scale tool to make a similar rectangle that has fd 20 as the first command.

   a. Click on the Scale tool. The Scale tool works much like the Change Shape tool, except that as you drag a corner side, the figure keeps the same shape. Figures that have the same shape—*similar* figures—are in proportion (like figures that are duplicated, reduced, or enlarged by a copy machine). Thus if the original rectangle's length is twice its width, all similar rectangles made with the Scale tool will have a length twice their width.

   b. With the Scale tool, drag a corner or side. The commands change automatically. Watch the forward commands and stop when they are twice as large as the original ones.

   c. Release the mouse button when you are done.

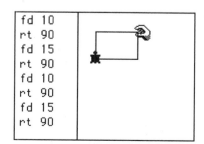

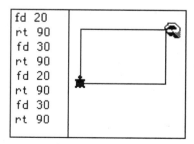

Notice which commands are changing when you use the Scale tool. The turn commands are not changing, but the forward commands do change. As they change, they stay in proportion to the original.

2. Predict what the commands for another similar rectangle would be if the first command were fd 30.

3. Use the Scale tool to make this similar rectangle and check your prediction.

Note: To use the Scale tool, the Command window must have a pattern of only fd and rt or lt commands.

With the number of decimal places set at zero, you can generate only similar polygons whose side measures are whole numbers. To generate additional similar figures, select **Decimal Places** in the **Options** menu and increase the number of decimal places shown. But note that, due to rounding, the numbers that show in the Command Center may not be precise when one or more decimal places are used.

Open *Geo-Logo* and choose **[Similar Houses]**. (Or, if you are already in *Geo-Logo*, choose **Change Activity** from the **File** menu, then single-click on the desired activity on the *Geo-Logo* activities screen.) A dialogue box appears with directions. Click **[OK]**.

How to Make Similar Houses

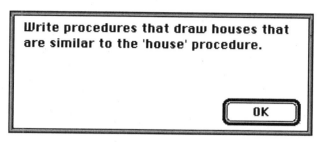

> **Write procedures that draw houses that are similar to the 'house' procedure.**
>
> OK

As the activity opens, the house procedure runs and the turtle draws what is referred to as the "original" house.

For this activity, students are working Student Sheet 24, Similar Houses, which asks them to draw four houses similar to the original one, working from a given side length.

1. Predict what the commands would be to make a similar house whose sides are double (or twice as long as) the original.

2. Enter your predicted commands in the Command Center.

   a. Click to the right of fd 8 in the Command Center.

   b. Press the **<delete>** key to delete the 8.

   c. Enter 16.

   d. Continue, doubling the input to each fd command.

   e. Press the **<return>** key when you are done to run your revised commands.

3. If you want the turtle to start in a new place, drag the turtle to the desired starting position before you run the procedure. A jumpto command will be generated in the Command Center. For example, you could drag the turtle to the lower left-hand corner to make room for a big house.

How to Use the Scale Tool for Help

If students are having trouble finding the pattern for drawing similar houses, they can use the Scale tool to create one automatically. Start with the original house.

1. Use the Scale tool to make a house whose sides are double (twice as large as) the original.

   a. Click on the Scale tool.

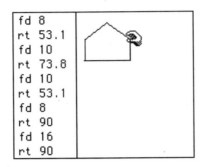 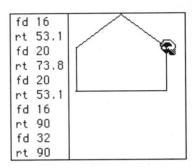

```
fd 8
rt 53.1
fd 10
rt 73.8
fd 10
rt 53.1
fd 8
rt 90
fd 16
rt 90
```

```
fd 16
rt 53.1
fd 20
rt 73.8
fd 20
rt 53.1
fd 16
rt 90
fd 32
rt 90
```

   b. Drag a corner or side. Watch the fd commands and stop when they are twice as large as the originals.

   c. Release the mouse button when you are done.

By comparing the original commands (in the Teach window or on Student Sheet 24) with the new commands in the Command Center, students should discover the pattern of doubling the number in each fd command.

How to Use the Scale Tool to Check Your Procedure

After you have entered your own commands in the Command Center, you can check to see whether your house procedure is actually similar to the original house.

1. Use the Scale Tool to scale your house to the size of the original house.

   a. Click on the Scale tool. Click on a corner of the house with the hand cursor.

   b. Drag a corner or side. Watch the fd commands as they change. Continue dragging until they match the measures of the original house. If the new house cannot be dragged to match the original exactly, your procedure is *not* similar to the original.

   In this example, the scaled-down commands do match the original commands, indicating that your house is similar.

   c. Release the mouse button when you are done.

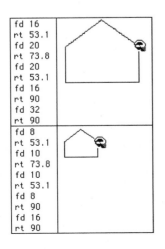

```
fd 16
rt 53.1
fd 20
rt 73.8
fd 20
rt 53.1
fd 16
rt 90
fd 32
rt 90
```

```
fd 8
rt 53.1
fd 10
rt 73.8
fd 10
rt 53.1
fd 8
rt 90
fd 16
rt 90
```

Free Explore

The Free Explore activity is available for you to use to extend and enhance activities in the unit and as an environment to further explore mathematics with *Geo-Logo*. Develop your own ideas and projects.

In Free Explore, the turtle responds to all *Geo-Logo* commands and all tools are available from the tool window.

Help

Assistance is available as you work with *Geo-Logo* activities. From the **Help** menu, choose any of the following:

Windows provides information on *Geo-Logo*'s three main windows: Command Center, Drawing, and Teach.

Vocabulary provides lists of *Geo-Logo*'s commands and examples.

Tools provides information on *Geo-Logo*'s tools (represented on the Tool bar as icons).

Directions provides instructions for the present activity.

Hints gives a series of hints on the present activity, one at a time. It is dimmed when there are no available hints.

Commands

The following commands are available in *Geo-Logo*. Some were previously introduced with activities in which they are useful. To see all of *Geo-Logo*'s commands, select **Vocabulary** from the **Help** menu.

Command	What It Means	What It Does
bk 10	back 10	Moves the turtle back 10 steps (or any number you specify). The turtle leaves a path if its pen is down. See also pd, pu.
ct	clear text	Clears, or erases, all the text in the Print window.
eraseall	erase all	Erases all the commands in the Command Center.
fd 50	forward 50	Moves the turtle forward 50 steps (or any number you specify). The turtle leaves a path if its pen is down. See also pd, pu.
fill	fill	Fills a closed shape or the entire Drawing window with the current turtle's color, starting at the current turtle's position. If the turtle's pen is over a path, only that path is filled. To fill a shape, use pu, then fd and rt or lt to move inside the shape. Use setc to set the color, then fill. Be sure that if the shape you want to fill is on a grid, you turn off the grid before filling. Use the hp command to hide points if you want to use jumpto or motions with fill.

`hp`	hide points	Hides points (they become invisible).
`ht`	hide turtle	Hides the current turtle (it becomes invisible).
`jumpto [-30 10]`	jump to	Moves the turtle to the point whose co ordinates are in the brackets without drawing.
`jumpto A`	jump to	Moves the turtle to point A without drawing. The letter must first be defined as a point.
`jt`	jump to	Abbreviation for `jumpto`.
`lt 120`	left 120	Turn the turtle left 120° (or any number you specify).
`make-points [a [10 5] b [20 40]] pd`		Makes points at the specified coordinates and shows them on the Drawing window. See the Make Points tool, which automatically generates this command.
`pd`	pen down	Puts the turtle's pen down so that when it moves, it draws a path.
`print [My drawing]`		Prints whatever is named in brackets, or a word or number in brackets, in the Print window. Can be used as a calculator. If you type `pr 85 + 15`, *Geo-Logo* will print `100`. Can be abbreviated `pr`.
`print colors`		Prints a list of colors to use with the `setc` command.
`pu`	pen up	Puts the turtle's pen up so that when it moves, it does not draw a path.
`recycle`		Cleans up and reorganizes available memory.
`repeat 4 [fd 10 rt 90]`		Repeats the commands in the list the specified number of times; in this example, 4 times. (The list is whatever appears in brackets.)
`rt 45`	right 45	Turns the turtle right 45° (or any number you specify).
`setc black`	set color	Sets the turtle's color; this affects the color of the turtle and the color for drawing and filling. The color names are:

```
white  black  gray  gray2  yellow
orange  red  pink  violet  blue
blue2 green  green2  brown
brown2  gray3
```

setxy [50 89]	set x, y	Moves the turtle to the point whose coordinates are in the brackets. If the pen is down, draws.
setxy A		Moves the turtle to point A. The letter must first be defined as a point. If the pen is down, draws.
sp	show points	Shows points and labels.
st	show turtle	Shows the turtle.

Menus

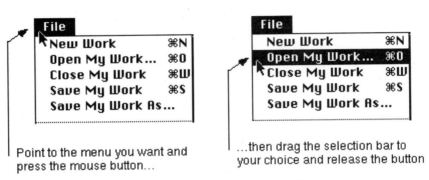

Point to the menu you want and press the mouse button...

...then drag the selection bar to your choice and release the button

Use the menus with the mouse, as illustrated. Some menu choices are also available from the keyboard. On the menu, the ⌘N indicates that, instead of selecting the choice from the menu, you could hold down the Command key (with the ⌘ and   symbols), then press the **<N>** key.

A menu choice may be dimmed, indicating it is not available in a particular situation.

Geo-Logo's Menus

The **File** menu deals with documents and quitting.

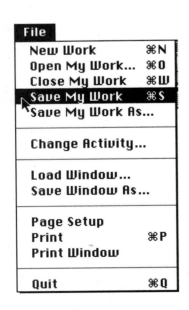

New Work starts a new document.

Open My Work... opens previously saved work.

Close My Work closes present work.

Save My Work saves the work.

Save My Work As... saves the work with a new name or to a different disk or folder.

Change Activity... shows the activity screen to allow you to select a new activity.

Page Setup brings up a dialogue box to set up the page for printing.

Print prints a whole document.

Print Window prints only the active window (the last one clicked on).

Quit quits *Geo-Logo*.

The **Edit** menu contains choices to use when editing your work.

> **Undo** reverses the last action taken, such as Delete, Cut, Paste, Erase, or Erase All.
>
> **Cut** deletes the selected text and saves it to a space called the Clipboard.
>
> **Copy** places the selected text on the Clipboard.
>
> **Paste** places the contents of the Clipboard at the place where the cursor is flashing.
>
> **Clear** deletes the selected text but does not put it on the Clipboard. Works the same as pressing the **<delete>** key when text is selected.
>
> **Stopall** stops any command or procedure that is running.

The **Font** menu is used to change the appearance of text. The change applies to the active window (the Command Center, Teach, Print, or Notes window).

> The first names are choices of typeface.
>
> **Size** and **Style** have additional choices; pull down to see the choices and then move to the right to select a choice. See the example for **Style**.
>
> **All Large** changes all text in all windows to a large-size font. This is useful for demonstrations. This selection toggles (changes back and forth) between **All Large** and **All Small**.

The **Windows** menu shows or hides the windows.

> If you Hide a window, such as the Drawing window, the menu item changes to **Show** followed by the name of the window: **Show Drawing**. You can also hide a window by clicking in the close box, the small square in the upper left-hand corner of the window.
>
> The **Show Print** choice opens the Print window and displays text generated by a print command from the Command Center. The **Show Notes** choice opens and hides the Notes window. You can use this window to enter and keep permanent notes.

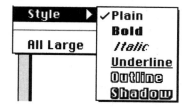

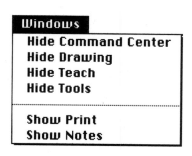

Options

Fast Turtle
Turn Rays

Decimal Places ▶
Scale Distance...

The **Options** menu allows you to customize *Geo-Logo*.

Fast Turtle turns the turtle quickly and so speeds up drawing. Usually in *Geo-Logo,* the turtle turns slowly to help students build images of the turns.

Turn Rays displays rays during turns to help students visualize the turn. After a turn command is entered, a ray is drawn to show the turtle's initial heading. Then as the turtle turns, another ray turns with it, showing the change in heading throughout the turn. A ray also marks every 30° of turn.

Decimal Places ▶
Scale Distance...

✓0
1
2
3
4
5
6

Decimal Places controls how many numbers after the decimal point are printed by certain commands and tools, such as the Ruler, Turtle Turner, Label Lengths, and Label Turns tools. For example, choose 1 to show tenths, 2 to show hundredths, and so on. If 0 (zero) is shown, the number is rounded to the nearest integer.

Scale Distance... controls units of measure for distance. A dialogue box enables you to change the unit of measure for a turtle step. If you enter 10 in the top left-hand box, the turtle would go forward 10 of the usual turtle steps when the command fd 1 was entered. If the [1 cm] or [1 inch] button is clicked, the proper number of steps is automatically entered, for example, 72 for 1 inch. If 1 inch were selected, the command fd 1 would move the turtle 1 inch on the screen. Note that this also encourages students to use fractions and decimals in issuing commands.

Help

Windows...
Vocabulary...
Tools...
Directions...   ⌘D
Hints...   ⌘H

The **Help** menu provides assistance.

Windows... provides information on *Geo-Logo's* three main windows: Command Center, Drawing, and Teach.

Vocabulary... provides a list of *Geo-Logo's* commands and examples of their use.

Tools... provides information on *Geo-Logo's* tools (represented on the Tool bar as icons).

Directions... provides instructions for the present activity.

Hints... gives a series of hints for the present activity, one at a time. It is dimmed when there are no available hints.

Only the most commonly used tools are available and displayed for each activity. All tools are available for Free Explore.

Tools

Teach

Teaches the turtle your procedure. Give the procedure a name. Enter its name in the Command Center to run it.

Ruler

Move the cursor to a point on the Drawing window. Click to find the length from the turtle to that point.

Line of Sight

Click and hold the mouse button to see a turtle turner show the turtle's heading.

Turtle Turner

Move the cursor on the Drawing window to turn the arrow. Click to show the turn command.

Draw Commands

Move the pointer and click to create a turn command. Then grab the turtle and pull it forward or back.

Change Shape

Click on a line segment or corner and drag it to its new location.

Grid

Puts a grid on the Drawing window, or removes it.

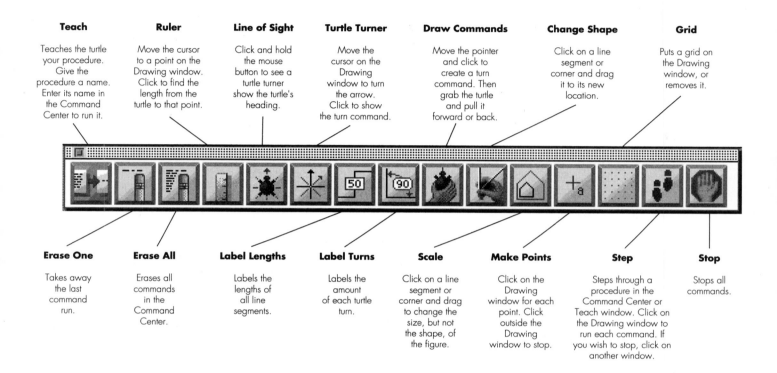

Erase One

Takes away the last command run.

Erase All

Erases all commands in the Command Center.

Label Lengths

Labels the lengths of all line segments.

Label Turns

Labels the amount of each turtle turn.

Scale

Click on a line segment or corner and drag to change the size, but not the shape, of the figure.

Make Points

Click on the Drawing window for each point. Click outside the Drawing window to stop.

Step

Steps through a procedure in the Command Center or Teach window. Click on the Drawing window to run each command. If you wish to stop, click on another window.

Stop

Stops all commands.

More on *Geo-Logo*'s Linked Windows

The *Geo-Logo* screen looks like this:

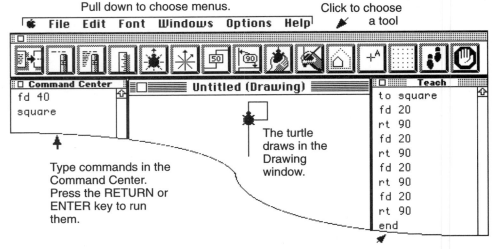

Pull down to choose menus.

Click to choose a tool

File Edit Font Windows Options Help

Type commands in the Command Center. Press the RETURN or ENTER key to run them.

The turtle draws in the Drawing window.

Defined procedures go in the Teach window.

Command Center

Type commands you wish the turtle to run immediately in the Command Center. Press the **<return>** key after each command. Make changes to commands directly in the Command Center; they are reflected automatically in the drawing when you press **<return>**. If you need to insert a line, hold down the Command key (with the ⌘ and   symbols on it), then press the **<L>** key.

Teach Window

When you have a sequence of commands you might wish to use again, you can define them as a procedure. Click on the Teach tool. A dialogue box appears, asking for a one-word name for this procedure. The name (with the word *to* in front of it) and the commands (with the word *end* added) are then placed in the Teach window on the right and named as a defined, or taught, procedure. You can then enter the name of the procedure as a new command.

If you change a procedure in the Teach window (for example, changing each fd 20 to fd 30 in the procedure square above), the change will show in the Drawing window as soon as you click out of the Teach window.

Drawing Window

You can also change a procedure by working directly in the Drawing window. The Draw Commands tool and the Change Shape tool, described in Tools on p. 149, allow you to change the geometric figure directly and see the effect on the *symbols* reflected immediately in the Command Center.

Trouble-Shooting

This section contains suggestions for how to correct errors, how to get back to what you want to be doing when you are somewhere else in the program, and what to do in some troubling situations.

If you are new to using the computer, you might also ask a computer coordinator or an experienced friend for help.

No *Geo-Logo* Icon to Open

■ Check that *Geo-Logo* Picturing Polygons has been installed on your computer by looking at a listing of the hard disk.

■ Open the folder labeled *Geo-Logo* Picturing Polygons by double-clicking on it.

■ Find the icon for the *Geo-Logo* Picturing Polygons application and double-click on it.

Nothing Happened After Double-Clicking on the *Geo-Logo* Icon

■ If you are sure you double-clicked correctly, wait a bit longer. *Geo-Logo* takes a while to open or load and nothing new will appear on the screen for a few seconds.

■ On the other hand, you may have double-clicked too slowly, or moved the mouse between your clicks. In that case, try again.

In Wrong Activity

■ Choose **Change Activity** from the **File** menu. Then click on the box of the activity you want.

Text Written in Wrong Area

■ Delete text.

■ Click cursor in the desired area or on the desired line and retype text; or select text and use **Cut** and **Paste** from the **Edit** menu to move text to desired area.

Out of Room in Command Center

- Continue to enter commands. Text will scroll up and old commands will still be there, but will be temporarily out of view. To scroll, click on the up or down arrows in the scroll bar along the right side of the window.

A Window Closed by Mistake

- Choose **Show Window** from the **Windows** menu.

Windows or Tools Dragged to a Different Position by Mistake

- Drag the window back into place by following these steps: Place the pointer arrow in the stripes of the title bar. Press and hold the button as you move the mouse. An outline of the window indicates the new location. Release the button and the window moves to that location.

I Clicked Somewhere and Now *Geo-Logo* Is Gone! What Happened?

You probably clicked in a part of the screen not used by *Geo-Logo,* and the computer therefore took you to another application, such as the Finder's "desktop."

- Click on a *Geo-Logo* window, if visible.
- Double-click on *Geo-Logo* Picturing Polygons again, or select it from the application menu at the right of the menu bar.

The Turtle Disappeared off the Screen. Why?

- If a command moves the turtle off the screen, write the opposite command to make it return. For example, if fd 500 sent the turtle off the screen, bk 500 will return it.

Many versions of Logo "wrap"—that is, when the turtle is sent off the top of the screen, it reappears from the bottom. *Geo-Logo* does not wrap when it is opened because students are learning to connect *Logo* commands to the geometric figures they draw. Each fd command in *Geo-Logo* makes a straight line segment.

How Do I Select a Section of Text?

In certain situations, you may wish to copy or delete a section or block of text.

- Point and click at one end of the text. Drag the mouse by holding down the mouse button as you move to the other end of the text. Release the mouse button. Use the **Edit** menu to **Copy, Cut,** and **Paste.**

System Error Message

Some difficulty with the *Geo-Logo* program or your computer caused the computer to stop functioning.

- Turn off the computer and repeat the steps to turn it on and start *Geo-Logo* again. Any work that you saved will still be available to open from your disk.

I Tried to Print and Nothing Happened

- Check that the printer is connected and turned on.

- When printers are not functioning properly, a system error may occur that causes the computer to "freeze." If there is no response from the keyboard or when moving or clicking with the mouse, you may have to turn off the computer and start over. See System Error Message above.

I Tried to Print the Drawing Window and Not Everything Printed

- Choose the Color/Grayscale option for printing.

- If your printer has no such option (for example, if it is an older black-and-white printer), you need to find a different printer to print graphics.

The turtle responds to *Geo-Logo* commands as a robot. If it does not understand a command, a dialogue box may appear with one of the following messages. Read the message, click on [OK] or press <return> from the keyboard, and correct the situation as needed.

Disk or directory full.

> The computer disk is full.
>
> ■ Use **Save My Work As...** to choose a different disk.

I don't know how to *name*.

> The program does not recognize the *name* command as written. Some possible problems are as follows:
>
> | | |
> |---|---|
> | fd50 | needs a space between fd and 50: fd 50 |
> | fdd 50 | extra *d* should be deleted |
> | mypicturje | misspelling to be corrected |

I don't know what to do with *name*.

> Either you gave too many inputs to a command, or no command at all.
>
> fd 50 30   needs only one number
>
> You may have left out a command.
>
> 5 + 16   change to print 5 + 16

I'm having trouble with the disk or drive.

> The disk might be write-protected, there is no disk in the drive, or some similar problem.
>
> ■ Use **Save My Work As...** to choose a different disk.

Name can only be used in a procedure.

> Certain commands, such as end, stop, and op (output), can't be used in the Command center.
>
> ■ Don't use that command if you don't need to.
>
> ■ Define the procedure in the Teach window.

Name does not like _name_ as input.

A command needs a certain type of input, which it didn't get from the command following it.

`fd fd 30`   omit one `fd` or put a number after the first one

`repeat [fd 30 rt 90]`

repeat needs two inputs: a number and a list. For example: `repeat 4 [fd 30 rt 90]`

Name is already used.

A procedure already exists with that name.

■ Use a different name.

Name needs more inputs.

Command _name_ needs an input, such as a number.

`fd`   needs how much to move forward. For example: `fd 30`

`rt`   needs how much to turn. For example: `rt 30`

Number too big

There are limits to numbers _Geo-Logo_ can use; it can use numbers up to 2147483647.

■ Don't exceed the limit.

Out of space

There is no free memory left in the computer.

■ Enter the command recycle to clean up and reorganize available memory.

■ Eliminate commands or procedures you don't need.

■ Save and start new work.

The maximum value for _name_ is _number._

The input is too high. For example:

The `maximum value for fd is 9999.`

■ Use a lower number.

The minimum value for _name_ is _number._

The input is too low a number. For example:

The `minimum value for fd is -9999.`

■ Use a higher number.

The *Geo-Logo* Picturing Polygons disk that you received with this unit contains the *Geo-Logo* program and a Read Me File. You may run the program directly from this disk, but it is better to put a copy of the program and the Read Me file on your hard disk and store the original disk for safekeeping. Putting a program on your hard disk is called installing it.

Note: *Geo-Logo* runs on a Macintosh II computer or above, with 4 MB of internal memory (RAM) and Apple System Software 7.0 or later. (*Geo-Logo* can run on a Macintosh with less internal memory, but the system software must be configured to use a minimum of memory.)

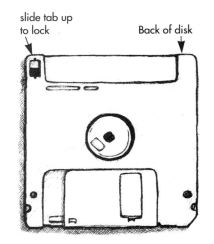

slide tab up to lock

Back of disk

To install the contents of the *Geo-Logo* Picturing Polygons disk on your hard drive, follow the instructions for your type of computer or these steps:

1. Lock the *Geo-Logo* Picturing Polygons program disk by sliding up the black tab on the back, so the hole is open.

   The *Geo-Logo* Picturing Polygons disk is your master copy. Locking the disk allows copying while protecting its contents.

2. Insert the *Geo-Logo* Picturing Polygons disk into the floppy disk drive.

3. Double-click on the icon of the *Geo-Logo* Picturing Polygons disk to open it.

4. Double-click on the Read Me file to open and read it for any recent changes in how to install or use *Geo-Logo*. Click in the close box after reading.

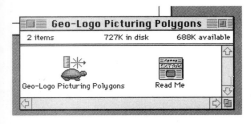

5. Click on and drag the *Geo-Logo* Picturing Polygons disk icon (the outline moves) to the hard disk icon until the hard disk icon is highlighted, then release the mouse button.

   A message appears indicating that the contents of the *Geo-Logo* Picturing Polygons disk are being copied to the hard disk. The copy is in a folder on the hard disk with the name *Geo-Logo* Picturing Polygons.

6. Eject the *Geo-Logo* Picturing Polygons disk by dragging it to the trash. Store the disk in a safe place.

7. If the hard disk window is not open on the desktop, open the hard disk by double-clicking on the icon.

When you open the hard disk, the hard disk window appears, showing you the contents of your hard disk. Among its contents is the folder labeled *Geo-Logo* Picturing Polygons, holding the contents of the *Geo-Logo* disk.

8. Double-click the *Geo-Logo* Picturing Polygons folder to select and open it.

When you open the *Geo-Logo* Picturing Polygons folder, the window contains the program and a Read Me file.

To select and run *Geo-Logo* Picturing Polygons, double-click on the program icon.

For ease at startup, you might create an alias for *Geo-Logo* Picturing Polygons by following these steps:

Optional

1. Select the program icon.

2. Choose **Make Alias** from the **File** menu

The alias is connected to the original file that it represents, so when you open an alias, you are actually opening the original file. This alias can be moved to any location on the desktop.

3. Move the *Geo-Logo* Picturing Polygons alias out of the window to the desktop space under the hard disk icon.

For startup, double-click on the *Geo-Logo* Picturing Polygons alias instead of opening the *Geo-Logo* Picturing Polygons folder to start the program inside.

Saving Work on a Different Disk

For classroom management purposes, you might want to save student work on a disk other than the program disk or hard disk. Make sure that the save-to disk has been initialized (see instructions for your computer system).

1. Insert the save-to disk into the drive.

2. Choose **Save My Work As...** from the File menu.

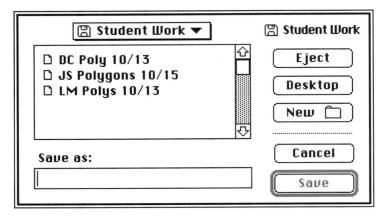

The name of the disk the computer is saving to is displayed in the dialogue box. To choose a different disk, click the **[Desktop]** button and double-click to choose and open a disk from the new menu.

3. Type a name for your work if you want to have a new or different name from the one it currently has.

4. Click on **[Save]**.

Deleting Copies of Student Work

As students no longer need previously saved work, you may want to delete their work (called *files*) from a disk. This cannot be accomplished from inside the *Geo-Logo* program. However, you can delete files from disks at any time by following the directions for your computer system.

Blackline Masters

General Resources for the Unit

_____ , 19 ____

Dear Family,

For the next few weeks, our class will be doing a unit called *Picturing Polygons*. In this unit, we will work with important mathematical ideas about two-dimensional objects by exploring polygons and coordinate geometry. Your child also will be exploring plastic shape pieces called Power Polygons and working on the computer to increase her or his knowledge of polygons and their properties.

As the unit unfolds, your child will work with:

- polygons with different numbers of sides, from 3 up to 10.
- polygons (particularly triangles and quadrilaterals) whose angles and sides follow certain rules.
- regular polygons (with all sides equal).
- similar polygons (of different sizes, but same shape).

When your child has assignments to work on at home, talk about them together and participate when asked. For example, how does your child find and draw a polygon that is "hidden" on a coordinate grid? When you and your child try to draw angles of a certain number of degrees—without any tools, just estimating—how close can both of you come? Is a square a rectangle? Is a rectangle a square? What *are* the rules for these polygons?

Look for opportunities to talk about shapes and angles with your child. Most of the angles in our rooms and on our furniture are right angles (90° angles). Why is this? Equilateral triangles have 60° angles. A diagonal across a square forms a 45° angle. If your child has any game boards or spinners, look at their designs. Do you see any of these angles? What other angles do you see?

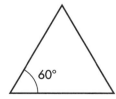

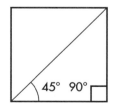

Above all, see how much fun shapes can be, and enjoy watching the growth of your child's understanding of geometry.

Sincerely,

Is It a Polygon?

1.

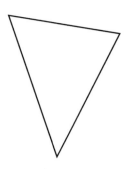

2.

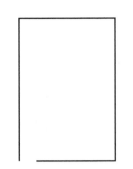

3.

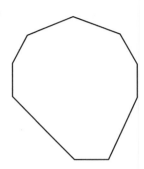

4.

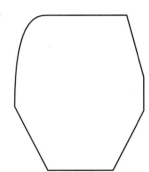

5.

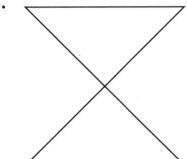

6.

7.

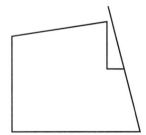

8.

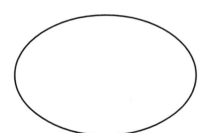

9.

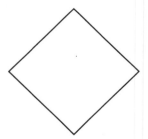

10.

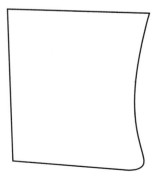

11.

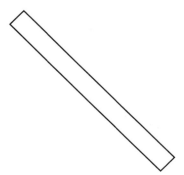

12.

Name _____ Date _____

Picasso's Polygons

How many polygons can
you find in this drawing?

Make your own picture
using **only** polygons.

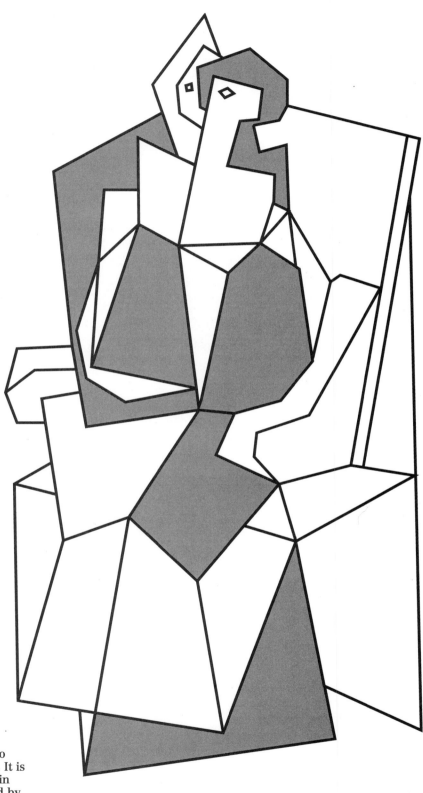

This figure is based on a pencil drawing by Pablo
Picasso, *Woman in an Armchair,* Biarritz (1918). It is
not a reproduction, but an illustration of the main
polygonal forms in the drawing. Adaptation used by
permission. Copyright © 1997 Estate of Pablo
Picasso/Artists Rights Society (ARS), New York.

Coordinate Grids

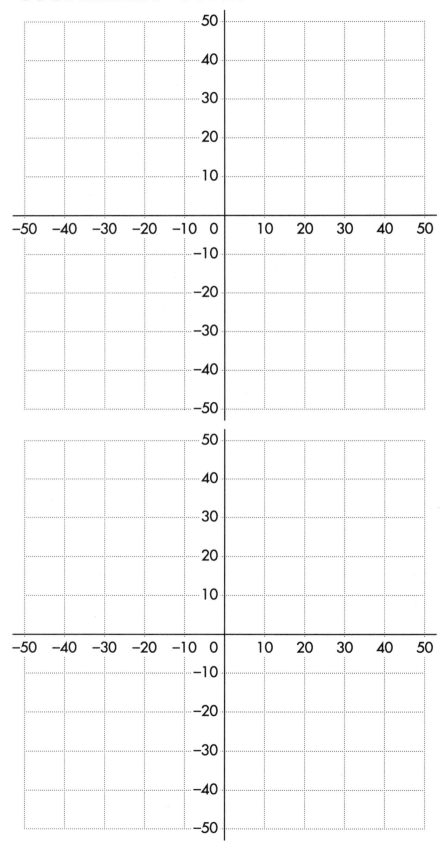

Hidden Polygon Pictures

Draw straight line segments
to each of these points in
order. Start at: (–20, –40)

(0, –10)

(30, –40)

(40, –30)

(10, 0)

(40, 20)

(–30, 30)

(–20, –40)

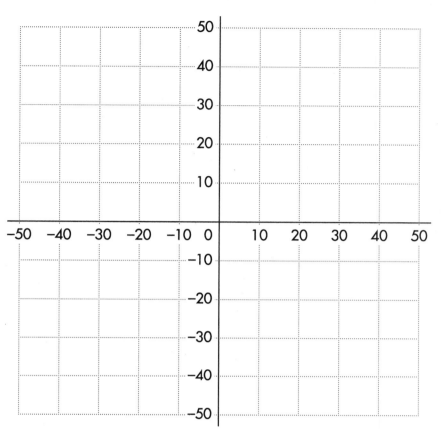

Draw straight line segments
to each of these points in
order. Start at: (–20, 20)

(–20, –10)

(–40, –10)

(–30, –30)

(20, –30)

(40, –10)

(20, –10)

(20, 10)

(–10, 10)

(–10, 20)

(–20, 20)

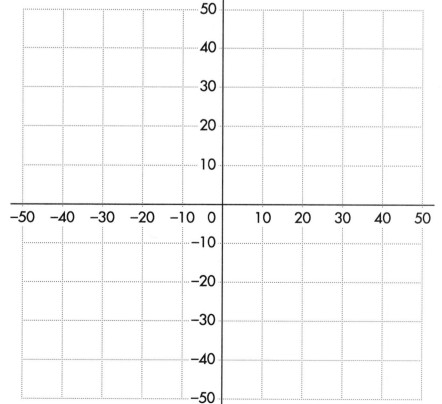

165

Types of Polygons

Fill in the second column of this chart with the name of the polygon for each number of sides.

Fill in the third column with related words or objects that use the same prefix (for example, tri- for triangles).

Write on a separate piece of paper any objects you see around you or know of that are examples of these polygons.

Number of sides	Name of polygon	Related words or objects
3		
4		
5		
6		
7		
8		
9		
10		
11		
12		

These are polygons.

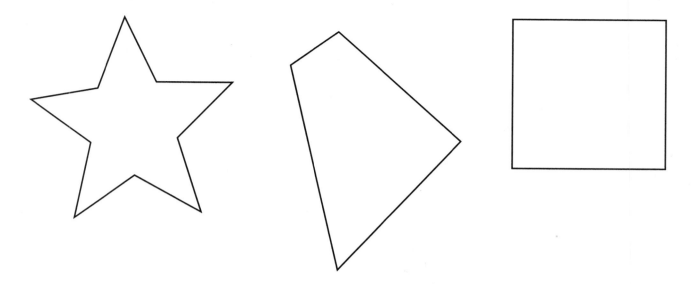

These are not polygons.

How to Get Started
- Open *Geo-Logo* with a double-click on the icon.
- Single-click the opening screen when instructed.
- Single-click an activity.

Geo-Logo Picturing Polygons

Investigations in Number, Data, and Space™
Picturing Polygons
Geo-Logo™

Click on this window to continue.
Authors & Programmers:
Douglas H. Clements & Julie Sarama
In collaboration with Michael T. Battista
© D. H. Clements, 1993; Logo core is © LCSI, 1993
Activities © 1995 Dale Seymour Publications

Geo-Logo Commands

setxy [10 -20] sets coordinates (10, –20)

jumpto [-10 15] moves turtle to (–10, 15)

      Can be abbreviated jt [–10 15]

fd 30	moves forward 30 turtle steps
bk 12	moves back 12 turtle steps
lt 90	turns left 90 degrees
rt 60	turns right 60 degrees

repeat 3 [fd 20 rt 90] repeats the commands in brackets 3 times

ht hides turtle (st shows turtle)

pu puts pen up (pd puts pen down)

pr 360 / 5 divides and prints answer in Print window

setc red changes drawing color to red

Tools

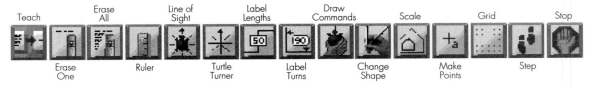

Teach   Erase All   Line of Sight   Label Lengths   Draw Commands   Scale   Grid   Stop

Erase One   Ruler   Turtle Turner   Label Turns   Change Shape   Make Points   Step

How to Get Help from *Geo-Logo*
- Choose a topic from the **Help** menu.

How to Open Saved Work
- Turn on the computer, open *Geo-Logo*.
- Choose **Open My Work**.
- Click the name of your work.
- Click [**Open**].

How to Save Your Work
- Choose **Save My Work** from the **File** menu.
- First time, type a name like this: DC+GW Rect 10/23
- Click [**Save**].

How to Finish
- Finish activity: Choose **Close My Work** from the **File** menu. STOP HERE if changing users.
- Finish *Geo-Logo*: Choose **Quit** from the **File** menu.
- Shut down and turn off the computer.

Name Date

Student Sheet 6

Is Every Three-Sided Polygon a Triangle?

Answer the following question, and draw sketches to illustrate your ideas.

Is every polygon with three sides a triangle? Why or why not?

Name Date

Student Sheet 7

Is Every Square a Rectangle? Is Every Rectangle a Square?

Answer the following questions, and draw sketches to illustrate your ideas.

Is every square a rectangle? Why or why not?

Is every rectangle a square? Why or why not?

How to Play Guess My Rule with Shapes

Materials: Deck of Guess My Rule cards

Two areas for grouping shapes according to whether or not they fit a rule—for example, a circle made of string or two different pieces of paper.

Players: 2

How to Play

1. The first player chooses a rule and gives a few examples, putting those shapes that fit the rule in one place (e.g., inside the circle) and those shapes that do not in the other place (e.g., outside the circle). The rule should focus on properties of geometric shapes, such as the shapes in the circle being all right triangles, all triangles with at least one right angle, or all quadrilaterals that are not squares.

2. The second player tries to guess the rule by placing a shape either inside or outside the circle, depending on whether the player thinks it fits the rule or not.

3. The first player says whether or not the placement is correct.

4. The second player uses this information to eliminate possibilities, devise new solutions, and revise earlier guesses of what the rule might be. Using this new information, the second player again tries to guess where a particular shape belongs.

5. Repeat steps 3 and 4. The second player can guess a rule if the player thinks he or she has found a solution. The first player says whether or not the rule is correct.

6. Play continues until the second player guesses the rule or there are no shapes left to place.

Can You Make These Triangles?

	Length of sides	Angles	Name of shape	Set xy points that make this triangle	Sketch of Power Polygons that make this triangle
tri1	your choice	1 right angle			
tri2	your choice	2 right angles			
tri3	3 equal sides	3 equal angles			
tri4	your choice	3 angles smaller than a right angle			
tri5	0 equal sides	1 angle larger than a right angle			

Do you think any of these triangles are impossible?
If so, pick one and write about how you could prove it is impossible.

Can You Make These Quadrilaterals?

	Length of sides	Angles	Number of parallel sides	Name of shape	Set xy points that make this quadrilateral.	Sketch of Power Polygons that make this quadrilateral
quad1	2 pairs of equal sides	4 right angles	2 pairs			
quad2	4 equal sides	0 right angles	2 pairs			
quad3	0 equal sides	4 right angles	your choice			
quad4	your choice	exactly 2 right angles	1 pair			
quad5	your choice	your choice	0 pairs			

Do you think any of these quadrilaterals are impossible?
If so, pick one and write about how you could prove it is impossible.

Find the Fourth Vertex

These are the coordinates of three points:

(0, 10)

(30, 10)

(20, –30)

Find a fourth point so that the four points form the vertices of a parallelogram.

How would you do this?

How many different points could be the fourth vertex?

What are the coordinates of these points?

Write your answers below.

Show the different possible parallelograms in different colors on the grid.

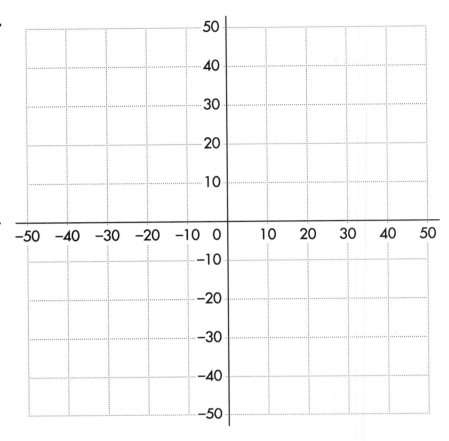

Some Shapes Fit Many Categories

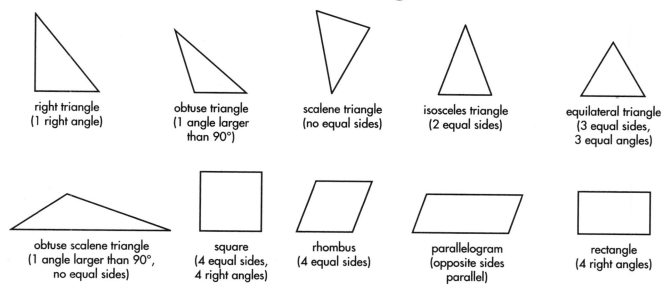

right triangle
(1 right angle)

obtuse triangle
(1 angle larger
than 90°)

scalene triangle
(no equal sides)

isosceles triangle
(2 equal sides)

equilateral triangle
(3 equal sides,
3 equal angles)

obtuse scalene triangle
(1 angle larger than 90°,
no equal sides)

square
(4 equal sides,
4 right angles)

rhombus
(4 equal sides)

parallelogram
(opposite sides
parallel)

rectangle
(4 right angles)

1. A square is a kind of rhombus. How can this be?

2. Name all the shapes above that are parallelograms. How can they be parallelograms and have other names as well?

3. An equilateral triangle is isosceles. How can this be?

4. Some obtuse triangles are scalene. Some obtuse triangles are isosceles. Sketch one or two examples of each.

5. Obtuse triangles cannot be equilateral. Explain why this is true.

What Shape Does It Draw?

Write the letter of each procedure in the shape that it draws.
Then write the procedure for the shape that is left.

to A	to B	to C	to D	to E
fd 20	fd 20	fd 20	fd 20	
rt 120	rt 60	rt 120	rt 90	
fd 20	fd 20	fd 20	fd 20	
rt 60	rt 150	rt 120	rt 90	
fd 20	fd 35	fd 20	fd 20	
rt 120	rt 150	rt 120	rt 90	
fd 20	end	end	fd 20	
rt 60			rt 90	
end			end	

1. 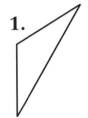 2.   3. 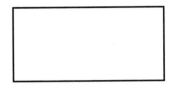 4. 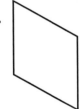 5.

Challenge 1
Write a procedure
to draw this shape.

Challenge 2
Pretend you are the turtle.
Draw the shape that this
procedure would draw.
Use a separate sheet of paper.

```
to drawme
lt 90
fd 5
rt 90
fd 10
lt 90
fd 15
rt 125
fd 35
rt 110
fd 35
rt 125
fd 15
lt 90
fd 10
rt 90
fd 5
end
```

Angles in the Power Polygons (page 1 of 2)

Label each angle with its size. Explain how you figured out the size of each angle.

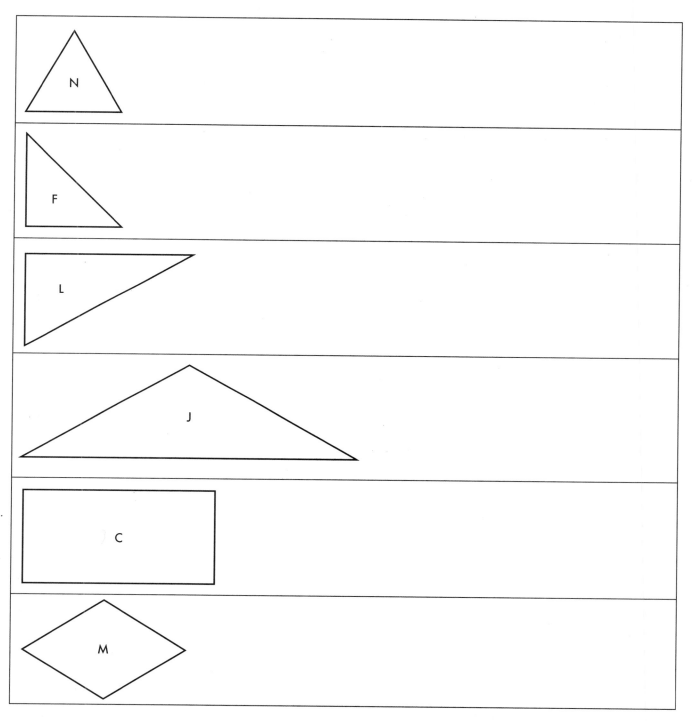

Angles in the Power Polygons (page 2 of 2)

Label each angle with its size. Explain how you figured out
the size of each angle.

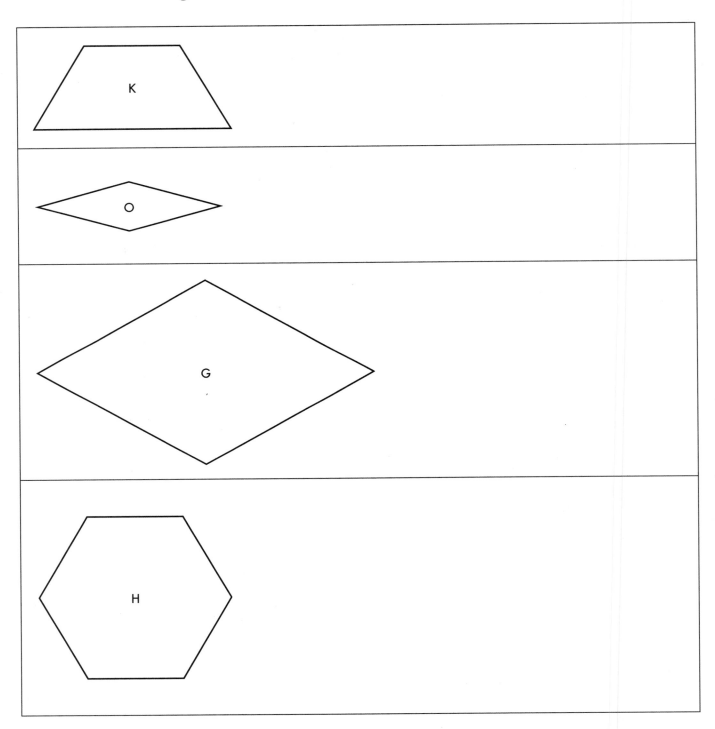

Estimating Angles

Draw the angle listed in each box. Then use your Turtle
Turner to check your accuracy. Challenge someone at home
to estimate the angles you draw in the last three boxes.

30°	45°
60°	90°
120°	Challenge angle for _____: (name of person)
Challenge angle for _____: (name of person)	Challenge angle for _____: (name of person)

Angles and Turns

Imagine the turtle walks straight up the left side of each triangle.
What **turn** does it make to head down the next side? What other
turns will it need to make as it walks around the shape?

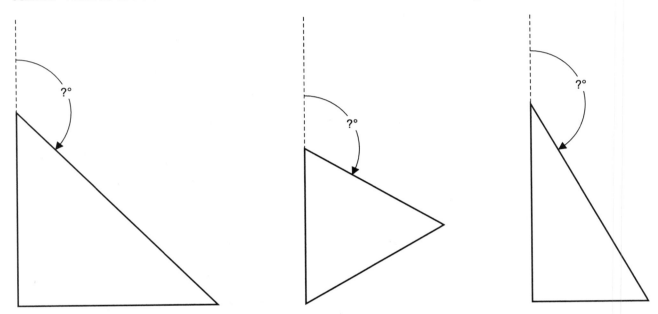

Figure out the number of degrees in the marked **angles** and
write them in. What sizes are the other angles in the triangles?

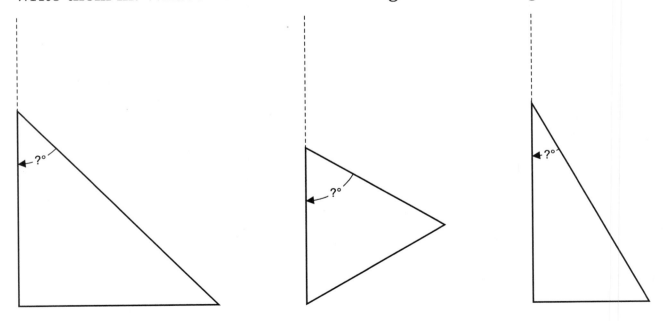

How are the angle size and the turn size related?

Name ______________________________ Date

Student Sheet 17

What Do You Know About 45° and 60° Angles?

Write about the two questions below. Include the following in your answers.

1. where you see 45° and 60° angles
2. the names of some polygons containing these angles
3. how 45° and 60° angles compare with other angles
4. how many of each angle you need to make a full circle or a straight line

What do you know about 45° angles?

____

____

____

____

____

What do you know about 60° angles?

____

____

____

____

____

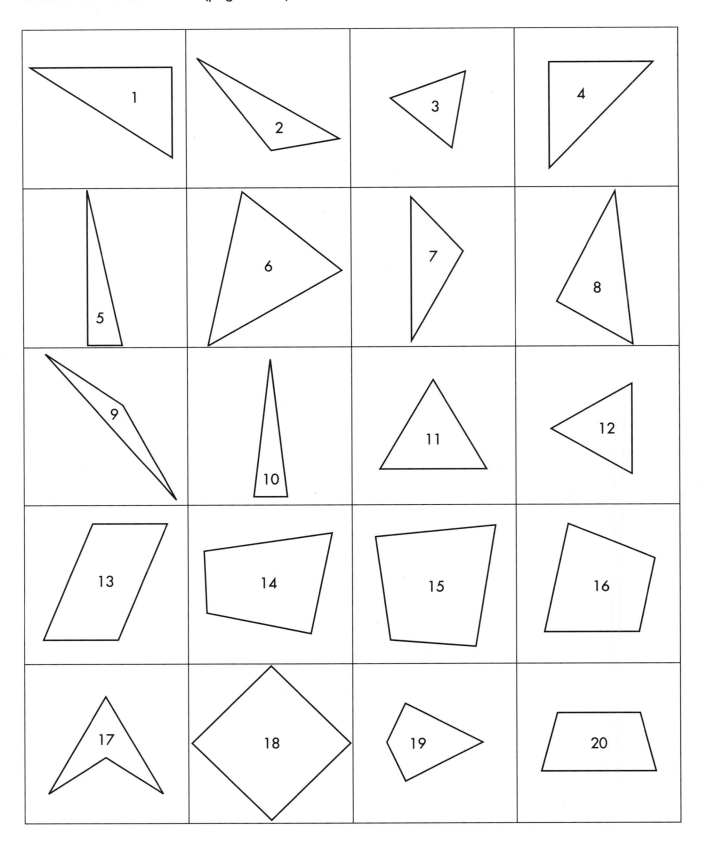

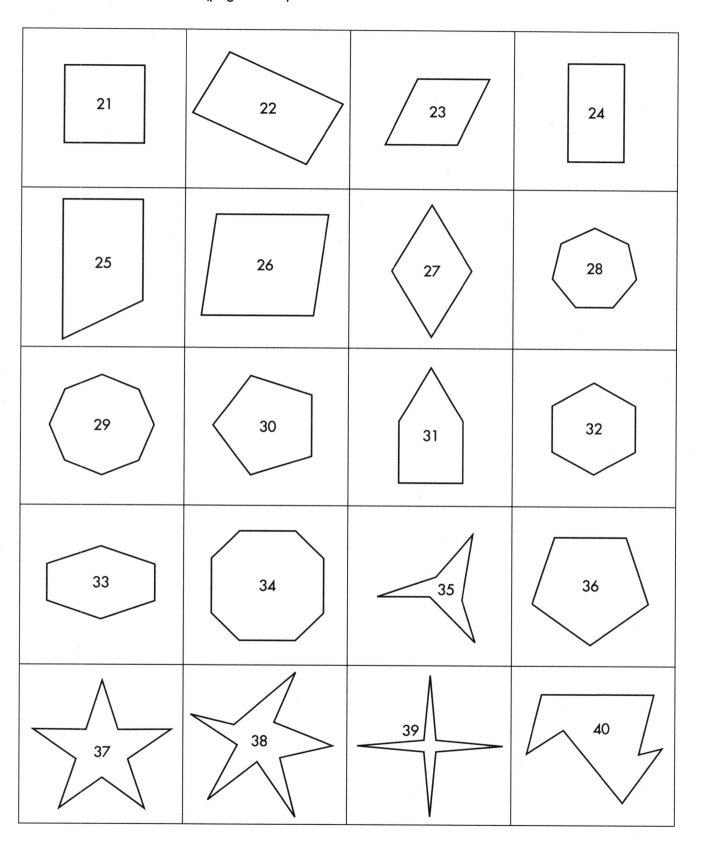

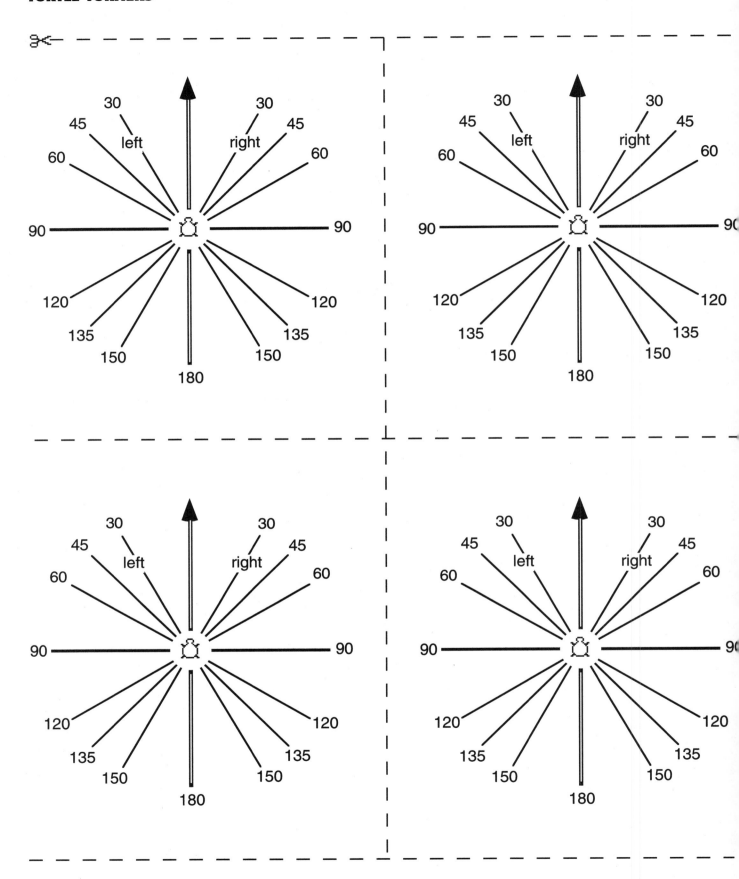

Name Date

Student Sheet 18

Total Turns and Angles (page 1 of 2)

Complete the chart below, working on or off the computer.
You may use drawings or Power Polygons to help.

Polygon	Number of sides, turns, angles	Length of sides (turtle steps)	Size of each turn (degrees)	Sum of turns	Size of each angle (degrees)	Sum of angles
equilateral triangle	3					
square	4					
regular pentagon	5					
regular hexagon	6					
regular heptagon (a challenge!)	7					
regular octagon	8					
regular nonagon	9					
regular decagon	10					

When you have finished the chart, write about any patterns you notice.

Total Turns and Angles (page 2 of 2)

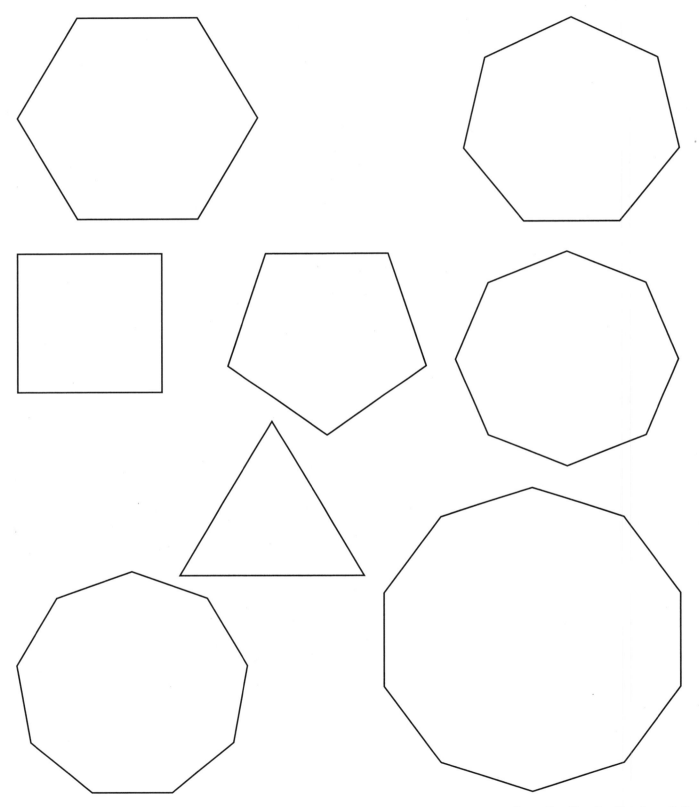

Which Are Regular Polygons?

Below are scale drawings of shapes made with Power Polygons.
Tell whether each shape is a regular polygon. Explain in writing
how you know.

1. Is this a regular polygon? _____ How do you know?

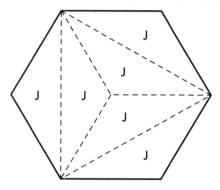

2. Is this a regular polygon? _____ How do you know?

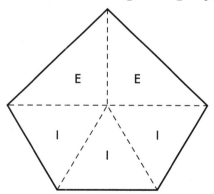

3. Is this a regular polygon? _____ How do you know?

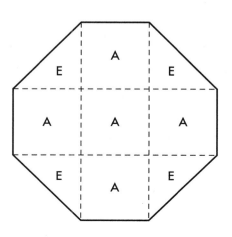

Polygons That Are Not Regular

Circle the group or groups you are to work with.

Group A: triangles and decagons
Group B: quadrilaterals and heptagons
Group C: pentagons and octagons
Group D: hexagons and nonagons

Draw at least two examples, with different-sized angles, of the nonregular polygons for your group(s). Measure the angles of each polygon with the Turtle Turner, and add them to find the sum.

Building Similar Shapes

Use Power Polygons. Build similar figures for each shape. Keep track of how many of each shape it takes to build the second larger shape, the third larger shape, and the fourth larger shape. **Predict** how many pieces it will take to build the tenth larger shape.

Shape	Number of pieces in similar (larger) shapes					
	1st	2nd	3rd	4th	5th	10th
B (square)						
C (rectangle)						
H (hexagon)						
J (obtuse isosceles triangle)						
K (trapezoid)						
M (rhombus)						
N (equilateral triangle)						
0 (thin rhombus)						

In the space below, show how you built the second trapezoid or hexagon by tracing around the shapes. On the back of this sheet, show how you built the third trapezoid or hexagon.

Name ____ Date

Student Sheet 22

Length of Sides Versus Area

Answer the following questions. Use sketches to help explain your answers.

1. When we double the length of the sides of a polygon, the area is 4 times as large. Why? Why not 2 times as large?

____

____

____

____

____

2. When we triple the length of the sides of a polygon, the area is 9 times as large. Why? Why not 3 times as large?

____

____

____

____

____

Similar Rectangles

Choose the *Geo-Logo* activity Similar Rectangles. The turtle draws the original rectangle in the Drawing window.

Write procedures to draw rectangles that are similar to the original rectangle. To check your procedures, use the Scale tool. See if the new rectangle stretches or shrinks to have exactly the same numbers as the original rectangle.

Original rectangle	Similar rectangle 1		Similar rectangle 2		Similar rectangle 3		Similar rectangle 4	
	Predict ☐	Corrected	Predict ☐	Corrected	Predict ☐	Corrected	Predict ☐	Corrected
fd 10	fd 20		fd 30		fd 16		fd 24	
rt 90	rt		rt		rt		rt	
fd 15	fd		fd		fd		fd	
rt 90	rt		rt		rt		rt	
fd 10	fd		fd		fd		fd	
rt 90	rt		rt		rt		rt	
fd 15	fd		fd		fd		fd	
rt 90	rt		rt		rt		rt	

Similar Houses

Choose the *Geo-Logo* activity Similar Houses. The turtle draws the original house in the Drawing window.

Write procedures to draw houses that are similar to the original house. Use the Scale tool to check your work.

Original house	Double		Triple		Half		Challenge	
	Predict ☐	Corrected	Predict ☐	Corrected	Predict ☐	Corrected	Predict ☐	Corrected
fd 8	fd 16		fd 24		fd 4		fd 20	
rt 53.1	rt		rt		rt		rt	
fd 10	fd		fd		fd		fd	
rt 73.8	rt		rt		rt		rt	
fd 10	fd		fd		fd		fd	
rt 53.1	rt		rt		rt		rt	
fd 8	fd		fd		fd		fd	
rt 90	rt		rt		rt		rt	
fd 16	fd		fd		fd		fd	
rt 90	rt		rt		rt		rt	

On the back of this sheet, explain how you found the commands for the triple-size house.

Investigation 3 • Sessions 5–6
Picturing Polygons

Drawing More Similar Rectangles

On pieces of one-centimeter graph paper, draw a rectangle
and then draw another that is similar to it.

Challenge someone at home to draw yet another rectangle
that is similar to the first two. Or provide one side of the
similar rectangle, and challenge someone to draw the rest,
making it similar to the original.

Take turns challenging each other. Experiment with both
wide and narrow rectangles, and make a few that are similar
to each.

These are regular polygons.

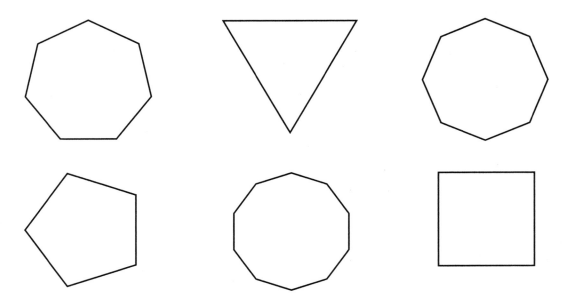

These are not regular polygons.

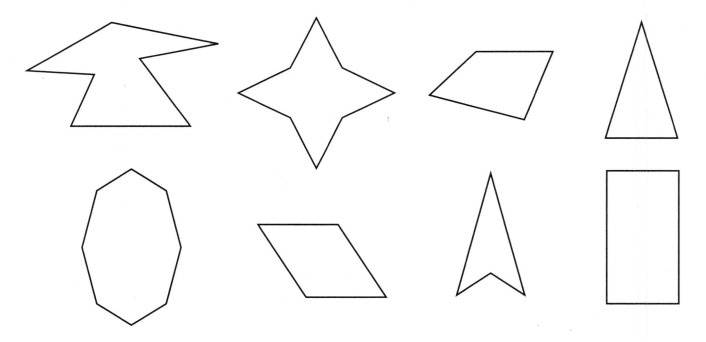

1	2	3	4	5	6	7	8	9	10
11	12	13	14	15	16	17	18	19	20
21	22	23	24	25	26	27	28	29	30
31	32	33	34	35	36	37	38	39	40
41	42	43	44	45	46	47	48	49	50
51	52	53	54	55	56	57	58	59	60
61	62	63	64	65	66	67	68	69	70
71	72	73	74	75	76	77	78	79	80
81	82	83	84	85	86	87	88	89	90
91	92	93	94	95	96	97	98	99	100

1	2	3	4	5	6	7	8	9	10
11	12	13	14	15	16	17	18	19	20
21	22	23	24	25	26	27	28	29	30
31	32	33	34	35	36	37	38	39	40
41	42	43	44	45	46	47	48	49	50
51	52	53	54	55	56	57	58	59	60
61	62	63	64	65	66	67	68	69	70
71	72	73	74	75	76	77	78	79	80
81	82	83	84	85	86	87	88	89	90
91	92	93	94	95	96	97	98	99	100
101	102	103	104	105	106	107	108	109	110
111	112	113	114	115	116	117	118	119	120
121	122	123	124	125	126	127	128	129	130
131	132	133	134	135	136	137	138	139	140
141	142	143	144	145	146	147	148	149	150
151	152	153	154	155	156	157	158	159	160
161	162	163	164	165	166	167	168	169	170
171	172	173	174	175	176	177	178	179	180
181	182	183	184	185	186	187	188	189	190
191	192	193	194	195	196	197	198	199	200
201	202	203	204	205	206	207	208	209	210
211	212	213	214	215	216	217	218	219	220
221	222	223	224	225	226	227	228	229	230
231	232	233	234	235	236	237	238	239	240
241	242	243	244	245	246	247	248	249	250
251	252	253	254	255	256	257	258	259	260
261	262	263	264	265	266	267	268	269	270
271	272	273	274	275	276	277	278	279	280
281	282	283	284	285	286	287	288	289	290
291	292	293	294	295	296	297	298	299	300

2	2	2	3
3	4	4	5
<u>6</u>	7	8	<u>9</u>
12	15	16	20
WILD CARD	WILD CARD	WILD CARD	WILD CARD

Ten-Minute Math
Picturing Polygons

FACTOR BINGO: MULTIPLICATION TABLE

	1	2	3	4	5	6	7	8	9	10	11	12
1	1	2	3	4	5	6	7	8	9	10	11	12
2	2	4	6	8	10	12	14	16	18	20	22	24
3	3	6	9	12	15	18	21	24	27	30	33	36
4	4	8	12	16	20	24	28	32	36	40	44	48
5	5	10	15	20	25	30	35	40	45	50	55	60
6	6	12	18	24	30	36	42	48	54	60	66	72
7	7	14	21	28	35	42	49	56	63	70	77	84
8	8	16	24	32	40	48	56	64	72	80	88	96
9	9	18	27	36	45	54	63	72	81	90	99	108
10	10	20	30	40	50	60	70	80	90	100	110	120
11	11	22	33	44	55	66	77	88	99	110	121	132
12	12	24	36	48	60	72	84	96	108	120	132	144

Ten-Minute Math
Picturing Polygons

100	180	200	60
98	32	72	150
240	144	324	225
448	396	330	450
WILD CARD	WILD CARD	WILD CARD	WILD CARD

Ten-Minute Math
Picturing Polygons

Practice Pages

This optional section provides homework ideas for teachers who want or need to give more homework than is assigned to accompany the activities in this unit. The problems included here provide additional practice in learning about number relationships and in solving computation and number problems. For number units, you may want to use some of these if your students need more work in these areas or if you want to assign daily homework. For other units, you can use these problems so that students can continue to work on developing number and computation sense while they are focusing on other mathematical content in class. We recommend that you introduce activities in class before assigning related problems for homework.

Close to 0   This game is introduced in the unit *Mathematical Thinking at Grade 5*. If your students are familiar with the game, you can simply send home the directions, score sheet, and Numeral Cards so that students can play at home. If your students have not played the game before, introduce it in class and have students play once or twice before sending it home. Students ready for more challenge can try the variation listed at the bottom of the sheet. You might have students do this activity two or three times for homework in this unit.

Solving Problems in Two Ways   Students explore different ways to solve computation problems in the units *Mathematical Thinking at Grade 5* and *Building on Numbers You Know*. Here, we provide two sheets of problems that students solve in two different ways. Problems may include addition, subtraction, multiplication, or division. Students record each way they solved the problem.

Counting Puzzles   In this kind of problem, introduced in the unit *Mathematical Thinking at Grade 5*, students are given a clue about a set of numbers. Students find three numbers that match the clue (there may be many numbers that would work). If necessary, you might distribute 300 charts for students to use. Provided here are two problem sheets and one 300 chart, which you can copy for use with the problem sheets. Because this activity is included in the curriculum only as homework, it is recommended that you briefly introduce it in class before students work on it at home.

Close to 0

Materials

■ One deck of Numeral Cards
■ Close to 0 Score Sheet for each player

Players: 1, 2, or 3

How to Play

1. Deal out eight Numeral Cards to each player.

2. Use any six cards to make two numbers. For example, a 6, a 5, and a 2 could make 652, 625, 526, 562, 256, or 265. Wild Cards can be used as any numeral. Try to make two numbers that, when subtracted, give you a difference that is close to 0.

3. Write these numbers and their difference on the Close to 0 Score Sheet. For example: 652 − 647 = 5. The difference is your score.

4. Put the cards you used in a discard pile. Keep the two cards you didn't use for the next round.

5. For the next round, deal six new cards to each player. Make two more numbers with a difference close to 0. When you run out of cards, mix up the discard pile and use them again.

6. After five rounds, total your scores. Lowest score wins.

Variation

Deal out ten Numeral Cards to each player. Each player uses eight cards to make two numbers that, when subtracted, give a difference that is close to 0.

Close to 0 Score Sheet

Game 1   Score

Round 1:   ___ ___ ___ – ___ ___ ___ = _____   _____

Round 2:   ___ ___ ___ – ___ ___ ___ = _____   _____

Round 3:   ___ ___ ___ – ___ ___ ___ = _____   _____

Round 4:   ___ ___ ___ – ___ ___ ___ = _____   _____

Round 5:   ___ ___ ___ – ___ ___ ___ = _____   _____

                                       TOTAL SCORE   _____

Game 2   Score

Round 1:   ___ ___ ___ – ___ ___ ___ = _____   _____

Round 2:   ___ ___ ___ – ___ ___ ___ = _____   _____

Round 3:   ___ ___ ___ – ___ ___ ___ = _____   _____

Round 4:   ___ ___ ___ – ___ ___ ___ = _____   _____

Round 5:   ___ ___ ___ – ___ ___ ___ = _____   _____

                                       TOTAL SCORE   _____

0	0	1	1
0	0	1	1
2	2	3	3
2	2	3	3

204

4	4	5	5
4	4	5	5
<u>6</u>	<u>6</u>	7	7
<u>6</u>	<u>6</u>	7	7

Practice Page
Picturing Polygons

8	8	9	9
8	8	9	9
WILD CARD	**WILD CARD**		
WILD CARD	**WILD CARD**		

Practice Page
Picturing Polygons

Practice Page A

Solve this problem in two different ways, and write about how you solved it:

375 − 268 =

Here is the first way I solved it:

Here is the second way I solved it:

Practice Page B

Solve this problem in two different ways, and write about how you solved it:

540 + 623 =

Here is the first way I solved it:

Here is the second way I solved it:

1	2	3	4	5	6	7	8	9	10
11	12	13	14	15	16	17	18	19	20
21	22	23	24	25	26	27	28	29	30
31	32	33	34	35	36	37	38	39	40
41	42	43	44	45	46	47	48	49	50
51	52	53	54	55	56	57	58	59	60
61	62	63	64	65	66	67	68	69	70
71	72	73	74	75	76	77	78	79	80
81	82	83	84	85	86	87	88	89	90
91	92	93	94	95	96	97	98	99	100
101	102	103	104	105	106	107	108	109	110
111	112	113	114	115	116	117	118	119	120
121	122	123	124	125	126	127	128	129	130
131	132	133	134	135	136	137	138	139	140
141	142	143	144	145	146	147	148	149	150
151	152	153	154	155	156	157	158	159	160
161	162	163	164	165	166	167	168	169	170
171	172	173	174	175	176	177	178	179	180
181	182	183	184	185	186	187	188	189	190
191	192	193	194	195	196	197	198	199	200
201	202	203	204	205	206	207	208	209	210
211	212	213	214	215	216	217	218	219	220
221	222	223	224	225	226	227	228	229	230
231	232	233	234	235	236	237	238	239	240
241	242	243	244	245	246	247	248	249	250
251	252	253	254	255	256	257	258	259	260
261	262	263	264	265	266	267	268	269	270
271	272	273	274	275	276	277	278	279	280
281	282	283	284	285	286	287	288	289	290
291	292	293	294	295	296	297	298	299	300

Practice Page
Picturing Polygons

Practice Page C

Find three numbers that fit each clue.

1. If you count by this number, you will say 36, but you will not say 40.

2. If you count by this number, you will say 120, but you will not say 130.

3. If you count by this number, you will say 66, but you will not say 77.

Practice Page D

Find three numbers that fit each clue.

1. If you count by this number, you will say 56, but you will not say 58.

2. If you count by this number, you will say 300, but you will not say 250.

3. If you count by this number, you will say 100, but you will not say 60.